ENDPAPER

A radiant Amelia Earhart waves to a cheering throng of admirers at Oakland Municipal Airport in 1935, as she completes the first solo flight from Hawaii to the American mainland. Perhaps the most accomplished of all early women pilots, she was feted frequently with welcomes like this one, depicted in this specially commissioned *gouache* rendering by distinguished artist Paul Lengellé.

WOMEN ALOFT

Other Publications:

THE CIVIL WAR
PLANET EARTH
COLLECTOR'S LIBRARY OF THE CIVIL WAR
LIBRARY OF HEALTH
CLASSICS OF THE OLD WEST
THE GOOD COOK
THE SEAFARERS
THE ENCYCLOPEDIA OF COLLECTIBLES
THE GREAT CITIES
WORLD WAR II
HOME REPAIR AND IMPROVEMENT
THE WORLD'S WILD PLACES
THE TIME-LIFE LIBRARY OF BOATING
HUMAN BEHAVIOR
THE ART OF SEWING
THE OLD WEST
THE EMERGENCE OF MAN
THE AMERICAN WILDERNESS
THE TIME-LIFE ENCYCLOPEDIA OF GARDENING
LIFE LIBRARY OF PHOTOGRAPHY
THIS FABULOUS CENTURY
FOODS OF THE WORLD
TIME-LIFE LIBRARY OF AMERICA
TIME-LIFE LIBRARY OF ART
GREAT AGES OF MAN
LIFE SCIENCE LIBRARY
THE LIFE HISTORY OF THE UNITED STATES
TIME READING PROGRAM
LIFE NATURE LIBRARY
LIFE WORLD LIBRARY

FAMILY LIBRARY:
HOW THINGS WORK IN YOUR HOME
THE TIME-LIFE BOOK OF THE FAMILY CAR
THE TIME-LIFE FAMILY LEGAL GUIDE
THE TIME-LIFE BOOK OF FAMILY FINANCE

This volume is one of a series that traces the adventure and science of aviation, from the earliest manned balloon ascension through the era of jet flight.

WOMEN ALOFT

by Valerie Moolman

AND THE EDITORS OF TIME-LIFE BOOKS

TIME-LIFE BOOKS, ALEXANDRIA, VIRGINIA

THE AUTHOR

Valerie Moolman, a New York-based writer, is the author of numerous books and documentary film scripts. A former text editor at Time-Life Books, she also wrote *The Road to Kitty Hawk* for this series.

THE CONSULTANT for Women Aloft

Claudia M. Oakes was appointed Assistant Curator of Aeronautics at the National Air and Space Museum, Washington, in 1975. She is the author of the book *United States Women in Aviation through World War I.*

THE CONSULTANTS for The Epic of Flight

Charles Harvard Gibbs-Smith was Research Fellow at the Science Museum, London, and a Keeper-Emeritus of the Victoria and Albert Museum, London. He wrote or edited some 20 books and numerous articles on aeronautical history. In 1978 he was the first Lindbergh Professor of Aerospace History at the National Air and Space Museum, Smithsonian Institution, Washington.

Dr. Hidemasa Kimura, honorary professor at Nippon University, Tokyo, is the author of numerous books on the history of aviation and is a widely known authority on aeronautical engineering and aircraft design. One plane that he designed established a world distance record in 1938.

For information about any Time-Life book, please write:
Reader Information
Time-Life Books
541 North Fairbanks Court
Chicago, Illinois 60611

Second Printing. Revised 1983.
Printed in U.S.A.
Published simultaneously in Canada.
School and library distribution by Silver Burdett Company, Morristown, New Jersey.

Library of Congress Cataloguing in Publication Data
Moolman, Valerie.
Women aloft.
(Epic of flight)
Bibliography: p.
Includes index.
1. Women in aeronautics. 2. Women air pilots.
I. Time-Life Books. II. Title. III. Series.
TL553.M78 629.13'088042 80-29475
ISBN 0-8094-3287-0
ISBN 0-8094-3288-9 (lib. bdg.)
ISBN 0-8094-3289-7 (retail ed.)

CONTENTS

Hats off to the ladies

No sooner had the first intrepid male aviators safely returned to earth, it seemed, than women, too, were smitten by an urge to fly. From mere spectators they became willing passengers and finally pilots in their own right, pitting their skill and daring against the hazards of the air and the skepticism of their male counterparts. In doing so, they enlarged the traditional bounds of a woman's world, won for their sex a new sense of competence and achievement, and contributed handsomely to the progress of aviation.

But recognition of their abilities did not come easily. "Men do not believe us capable," the famed aviator Amelia Earhart once remarked to a friend. "Because we are women, seldom are we trusted to do an efficient job." Indeed, old attitudes died hard: When Charles Lindbergh visited the Soviet Union in 1938 with his wife, Anne—herself a pilot and gifted proponent of aviation—he was astonished to discover both men and women flying in the Soviet Air Force. "I don't see how it can work very well," he later confided to his diary. "After all, there is a God-made difference between men and women that even the Soviet Union can't eradicate."

Such conventional wisdom made it difficult for women to raise money for the up-to-date equipment they needed to compete on an equal basis with men. Yet compete they did, and often they triumphed handily despite the odds.

Ruth Law *(left)*, whose 590-mile flight from Chicago to Hornell, New York, set a new nonstop distance record in 1916, exemplified the resourcefulness and grit demanded of any woman who wanted to fly. And when she addressed the Aero Club of America after completing her historic journey, her plain-spoken words testified to a universal human motivation that was unaffected by gender: "My flight was done with no expectation of reward," she declared, "just purely for the love of accomplishment."

Waving their caps in tribute, soldiers welcome Ruth Law to Governors Island in New York on November 20, 1916, after her record-breaking long-distance flight.

1

The first hurdle: proving they were fit to fly

Harriet Quimby (right) and Matilde Moisant, America's first two licensed women pilots, appear the essence of frilly femininity in this photograph taken shortly after they learned to fly in 1911 at a school owned by Matilde Moisant's brother.

On a balmy June day in 1784, Elisabeth Thible of Lyons, France—wearing a lace-trimmed dress and a feathered hat—climbed into the gondola of a hot-air balloon piloted by an artist named Fleurant and soared a mile above the rolling French countryside. She was so exhilarated she burst into song.

Ballooning was still in its infancy—the first manned flight had astonished Paris just seven months earlier—and Mme. Thible was the first woman to travel aloft. Her first flight was also her last, but she was followed by other women who flew first as passengers and later as professional balloonists making their living providing carnival audiences with a thrill.

The reception given these pioneer women fliers was curiously mixed. On the one hand, they were acclaimed as a novelty. On the other, they were distrusted as interlopers in an adventure that was thought to belong rightly to men. That equivocal attitude persisted throughout the 19th Century—and far into the age of powered flight. Women fliers were intriguing, imaginative and daring—but somehow they did not belong in the sky.

As late as 1911—more than one and a quarter centuries after Mme. Thible ascended—*The Detroit Free Press* was pondering the question "Ought women to aviate?" In the *American-Examiner,* the renowned British aviator Claude Grahame-White declared that the answer was categorically no. Women, he said, "are temperamentally unfitted" for flying because they are prone to panic. He had taught many women to fly, added Grahame-White, but regretted it: "When calamity overtakes my women pupils, as sooner or later I fear it will, I shall feel in a way responsible for their sudden decease."

Less partisan observers tirelessly debated women's place in the sky in respected aviation journals like the *Revue Aérienne* in France and *Flying* in the United States. Yet the discussion was in some ways superfluous—for women not only began flying shortly after men did but almost from the beginning made creative contributions to mankind's collective experience of flight. In their frail balloons they experimented with parachute drops, undertook scientific observations and tested themselves against the hazards of high altitudes. And with the coming of powered flight, they would do even more—exploring the limits of mechanical and human endurance as they strove to prove that they belonged in the sky. "Now and then," said famed long-distance flier

Amelia Earhart, "women should do for themselves what men have already done—and occasionally what men have not done—thereby establishing themselves as persons, and perhaps encouraging other women toward greater independence of thought and action."

Over the years, women would break long-distance and altitude records, fly as test pilots, probe untried routes, compete successfully with men in trials of endurance and speed, and even, in more than a few cases, teach men to fly. Routinely, they risked their lives, and sometimes they lost them. And when World War II came, they would demonstrate, in the words of the Air Surgeon of the United States Army Air Forces, that they were "adapted physically, mentally and psychologically" to the flying of high-speed military planes.

At no point was their path easy. To fly was to cajole men into giving them rides, to beg to be taught, to meet a wall of masculine resistance. Women not only had difficulty getting flight training comparable to that available to men but rarely were offered either financial backing or essential, productive flying jobs. And often they had to use equipment that was inferior to that used by men. "The reason why I went in for record-breaking and long-distance flying," recalled Jacqueline Cochran, "was simply that then, as now, it was only men who were allowed to act as test pilots and operate transport aircraft. Accordingly, I had only the choice between continuing to pilot light machines, which bored me and cost money, or getting hold of a fast, up-to-date aircraft in which I could try to break records. I might risk my skin that way, but I could probably earn a living."

Yet getting hold of a fast, up-to-date plane was all but impossible for most women—chiefly because high-powered machines were supposed to be too hot for women to fly. Flier Elinor Smith complained to an interviewer in 1930 that the public thought women could "only handle light planes." She added: "If you come in with a light plane nobody pays any attention to you. Heavy planes aren't really any harder to handle, but people think they are."

Attracting attention was important—for Elinor Smith, like many another woman flier, was plagued intermittently by financial problems throughout her career. Publicity was her lifeblood, and she courted it with such hazardous stunts as flying under all the East River bridges of Manhattan. To her critics she pointed out, truthfully enough, that she had not started flying for the sake of publicity but sought publicity so that she could continue to fly.

Other women fliers grubbed for a living writing newspaper stories, giving occasional lectures or radio talks or endorsing the brand of gas or oil they used. Viola Gentry, who in 1928 set the first women's solo endurance record, had paid for her lessons by working as a cashier in a Brooklyn cafeteria. With women's average annual income in 1940 standing at $850 and a pilot's license costing between $500 and $750, it is not surprising that many women fliers were pressed for money—or that many others came from backgrounds of wealth. Raymonde de

Mary Myers, one of the first American women balloonists to make a solo flight, stands in the basket of her balloon, the Aerial, on the cover of this 1883 pamphlet published by her husband, Carl Myers. Billed as Carlotta, the Lady Aeronaut, she became famous as an exhibition flier.

Drawn by the promise of an exhibition flight, an enthusiastic crowd gathers at Mary Myers' home in Frankfort, New York, where she and her husband ran a balloon factory and flying school. Their company produced the hydrogen gas balloon and sausage-shaped airship seen on the lawn.

Laroche, the first woman to obtain a license, was prominent in French society; in England in the 1920s, flying had strong social overtones and attracted women who were titled or wealthy or both; in the United States, a number of women fliers—Ruth Nichols and Laura Ingalls, for example—came out of fashionable private schools.

Yet most women fliers had no other resources than intelligence and enormous amounts of grit. Many of them came not from finishing schools but from farms. "To some young women with dreams of a wider world," recalled Elinor Smith, "there seemed to be two paths to follow, each with great romantic appeal. One led to Hollywood, the other to a career in the sky. For me there was only one path: I knew from the age of six that I wanted to fly." Louise Thaden, one of the great women racing pilots of the 1930s, came to aviation from a farm in Bentonville, Arkansas, via Wichita, Kansas, where she worked as a sales clerk for a coal company. In Wichita she hung about the local flying field and dreamed of that sense of "being master of my fate" that she felt only flying could bring. Evelyn "Bobbi" Trout, who emerged from the little town of Greenup in rural Illinois to set three women's endurance records in 1929, recalled being "drunk with flying" from the moment she saw a plane pass overhead.

What were they seeking, besides escape from the strictures of a too-narrow world? Some, like Louise Thaden, or like the fearless French flier Hélène Boucher, saw a chance to assert themselves in the air as

Persistent designer of peculiar planes

The first woman to design and build an airplane was not a pilot or engineer but an inventive stenographer named E. Lillian Todd. In 1906 she unveiled her first aircraft, a peculiar machine meant to be powered by its own movement through the air. Unfortunately, it owed more to wishful thinking than to the laws of physics; it never flew.

But Lillian Todd continued designing and proposed some practical ideas that later were widely adopted—such as planes that could be collapsed for easy transport. While none of her full-scale creations ever flew, she taught countless youngsters to build flyable models and founded the Junior Aero Club of America, through which hobbyists explored the wonders of the air.

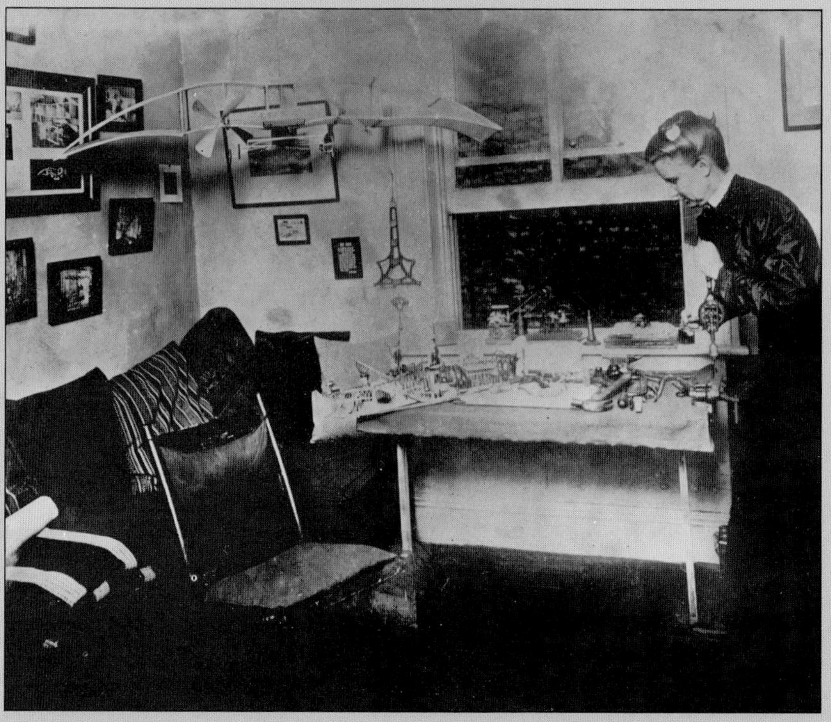

Lillian Todd works in her living-room studio, where the Junior Aero Club met.

At the 1908 Aeronautical Society exhibition in Morris Park, New Jersey, Miss Todd displays a model of one of her air-powered planes.

they could not on the ground. "It's the only profession where courage pays off and concrete results count for success," said Hélène Boucher.

Other women expressed an almost mystic sense about their profession, bringing their own special perceptions to the wonders and beauties of flight. Louisa Wise, an American balloonist of the mid-19th Century, spoke in wonder of the "solemn silence" of the open sky—which she said reminded her, as it did other early balloonists, of a soaring, empty cathedral. For later fliers, the silence was gone, but the wonder and mystery remained. On her flights with her famous husband, Anne Morrow Lindbergh became entranced by "the fundamental magic of flying"—which to her was like peering through "the ruffled surface of life, far down to that still permanent world below."

Some women flew for thrills: Flying, said the German racing pilot Thea Rasche, was "more thrilling than love for a man and far less dangerous." And many women flew at least partly because they wanted to assert their competitive rights as women. "Women are innately better pilots than men," insisted Louise Thaden—an attitude not uncommon among the women fliers of her day. To compete with men on equal terms was to strike a blow for female rights. When the acrobatic pilot Katherine Stinson became the first woman to do multiple loops, in 1915, she announced that having "equalled the greatest efforts of the male fliers," she now proposed to unveil a snap roll on top of a loop that would indisputably "put woman ahead of man in the most difficult of sciences." The stunt became a favorite with crowds. The successful attempt at a nonstop distance record made by Ruth Law in 1916 was a deliberate effort to trace the flight path of the previous record holder, Victor Carlstrom, and surpass him. Ruth Law was competing, recalled Amelia Earhart, against "the men of the day who, in training and equipment, were forging ahead of the few individual women struggling for a chance in the air."

Women paid for their competitiveness in a thousand large and small ways. When Germany's first woman flier, Melli Beese, took the test for her license in 1911, male colleagues at the flying field tried unsuccessfully to sabotage her by tampering with the plane's steering mechanism and partially emptying the gas tank; for a woman to fly, said one of them, "would take the glory away from us." Years later, the attitude had scarcely changed. Elinor Smith recalled that once when she was flying a dual-control Bellanca on a promotion tour in the late 1920s, her male copilot was so uneasy about being flown by a woman that he later claimed falsely to the press that it was he who had been piloting the plane. When racing flier Helen Richey was hired by Central Airlines in 1934 as the first woman pilot on a regularly scheduled airline, the pilots' union refused to accept her. Prevented by union pressure from flying in bad weather, she eventually resigned, saying she did not want to be a "fair-weather pilot."

The wonder is that women, in the male-dominated world of aviation, accomplished all that they did. The marks set by the Earharts and

Cochrans, the Thadens and Bouchers were not only challenges to pilots of both sexes, but inspirations to thousands of young women whose secret dream was to fly. Yet the real legacy of the great women fliers was something less tangible: By their very presence in the air, they helped to demystify—and even to domesticate—the awesome sky. They showed that it was a safe place not only for men but for women and for families—thus helping to prepare the way for the system of mass transportation that flying would one day become. As the aviation industry came to realize that women fliers had the capacity to dispel the terrors of flight, American aircraft companies hired virtually all the top women pilots of the day to tour the country extolling the safety and the future prospects of aviation. Famous figures such as Amelia Earhart wrote articles with such titles as "Why Are Women Afraid To Fly?" and "Shall You Let Your Daughter Fly?"

To speak for aviation was for most women fliers not an imposition but a privilege. In the reminiscences of the early women pilots there is a recurrent sense of the gratitude they felt at having belonged to flying when it was young. "Picture, if you can, what it meant for the first time, when all the world of aviation was young and fresh and untried," wrote the English balloonist Gertrude Bacon looking back on the marvels of her youth. "I have experienced something that can never be yours and can never be taken away from me—the rapture, the glory and the glamour of 'the very beginning.' "

An adventurous lady, her skirt securely tied against the wind, flies with Wilbur Wright on this 1909 cover of Femina, a popular French women's magazine that sponsored the Coupe Femina for female long-distance fliers. It was said that such pictures inspired a new fashion: the hobble skirt.

Women were piloting aircraft by 1798—France's Jeanne Labrosse made a solo balloon flight in that year—but the first woman to demonstrate she could be something other than a novelty in the air was Madeleine Sophie Blanchard, who first flew solo in 1805. Her husband, a noted French balloonist, remarked before dying of a lingering illness that he feared his wife would have no way to support herself. In fact, she became one of the most famous fliers of her day, attracting enormous crowds to her increasingly daring ascents. She was appointed official Aeronaut of the Empire by Napoleon and toured Europe. During an aerial fireworks display in 1819, she fell to her death, a victim of the public's appetite for the spectacular.

By 1834, no fewer than 22 women in Europe had piloted their own balloons. In England, Margaret Graham became almost a fixture of Victorian life during a 30-year career in which she popularized flying by taking passengers aloft for a fee. In the United States, Mary H. Myers set an astonishing world altitude record in 1886—soaring four miles above Franklin, Pennsylvania, without benefit of oxygen equipment. When, toward the end of the century, there was a surge of interest in sending up scientific research balloons, such women as England's Gertrude Bacon were among the researchers who made daring ascents. She overcame an initial fear of flying to take numerous photographs over land and water and to help her father test radiotelegraphy in the clouds. Not long afterward, in 1903, Cuban-born Aida de Acosta made one of the world's first powered flights, piloting a dirigible over Paris just months before the Wright brothers achieved powered heavier-than-air flight at Kitty Hawk.

Despite these and other exploits, flying remained a highly unusual activity for respectable women. In the early years of the 20th Century, even women flying as passengers sometimes tried to conceal their identities. When Violet Ridgeway, a leading socialite of Philadelphia, turned up for an early-morning balloon flight in October of 1909, she prudently gave her name as Miss Anna Brown and appeared heavily veiled.

It was in France, birthplace of the balloon and of Elisabeth Thible, that women earned their first grudging acceptance as pilots. Preeminent among them was Frenchwoman Raymonde de Laroche, who styled herself a baroness (her right to the title was in doubt, but she was generally known by it) and became the first woman to earn a pilot's license. She had by her own account painted portraits, created sculptures, performed on the stage, driven racing cars and made flights in balloons. Taken for an airplane ride, she exclaimed: "This is the way we humans are meant to travel!" She was delighted when Charles Voisin, one of the great air pioneers of France, offered to teach her to fly. On October 22, 1909, shortly after her twenty-third birthday, the baroness presented herself at the Châlons airfield where Charles and his brother, Gabriel, built and tested their planes.

The Voisin, which looked like a pair of box kites joined by a skeletal frame, was a one-seater like most of the planes of the time; the student took the controls while the instructor shouted guidance from the ground. Following Charles Voisin's instructions, the baroness settled herself in the open cockpit and taxied the length of the airfield to get the feel of the controls. She was not, she had been told, to make any attempt to take off.

After taxiing once around the airfield, the tyro pilot announced that she was ready to fly. A startled Charles Voisin, an admiring English reporter named Harry Harper and several incredulous mechanics watched her open up the throttle of the 50-horsepower engine, race the machine down the airstrip and rise to a height of about 15 feet. "Moving on a perfectly even keel," Harper wrote, the plane "skimmed through the air for a few hundred yards, and then settled gently and came taxiing back."

Contributing a paragraph to aviation history, the baroness passed her qualifying tests and on March 8, 1910, obtained from the Aero Club of France the first license issued to a woman anywhere in the world. Flying was ideal for women, she told the gathered reporters. "It does not rely so much on strength," she pointed out, "as on physical and mental coordination."

When a reporter suggested that flying was dangerous work for a woman, the baroness shrugged: "Most of us spread the hazards of a lifetime over a number of years. Others pack them into minutes or hours. In any case, what is to happen will happen."

Spectators and cavalrymen mill around the wreckage of the box-winged craft in which Raymonde de Laroche crashed at Rheims, France, in 1910. The plane was designed by her friend Charles Voisin.

Something happened to Raymonde de Laroche exactly four months later. She was competing in an endurance contest, the only woman flier in a galaxy of performers including such stars as Hubert Latham and Louis Blériot, when she encountered a phenomenon that was seldom experienced and barely understood at the time. As she was rounding a pylon, another airplane suddenly darted in front of her. "We saw her machine shudder and swerve," reported Harry Harper. "She made a gallant effort to regain control, but was too near the ground. The machine side-slipped and was completely wrecked as it hit not far from the pylon." She had been caught in the turbulent prop wash of the other plane.

She was alive, but barely. With serious head wounds, internal injuries, one arm and both legs broken, it seemed unlikely that she would recover, much less fly again.

But she mended and was racing again within two years. In 1913 she won the Coupe Femina, a cup established by the French magazine *Femina* to honor women fliers. "It may be that I shall tempt Fate once too often," she acknowledged to her friends. "But it is to the air that I have dedicated myself, and I fly always without the slightest fear." Ironically, she was not at the controls but flying simply as a passenger when she was killed in the crash of an experimental plane in 1919.

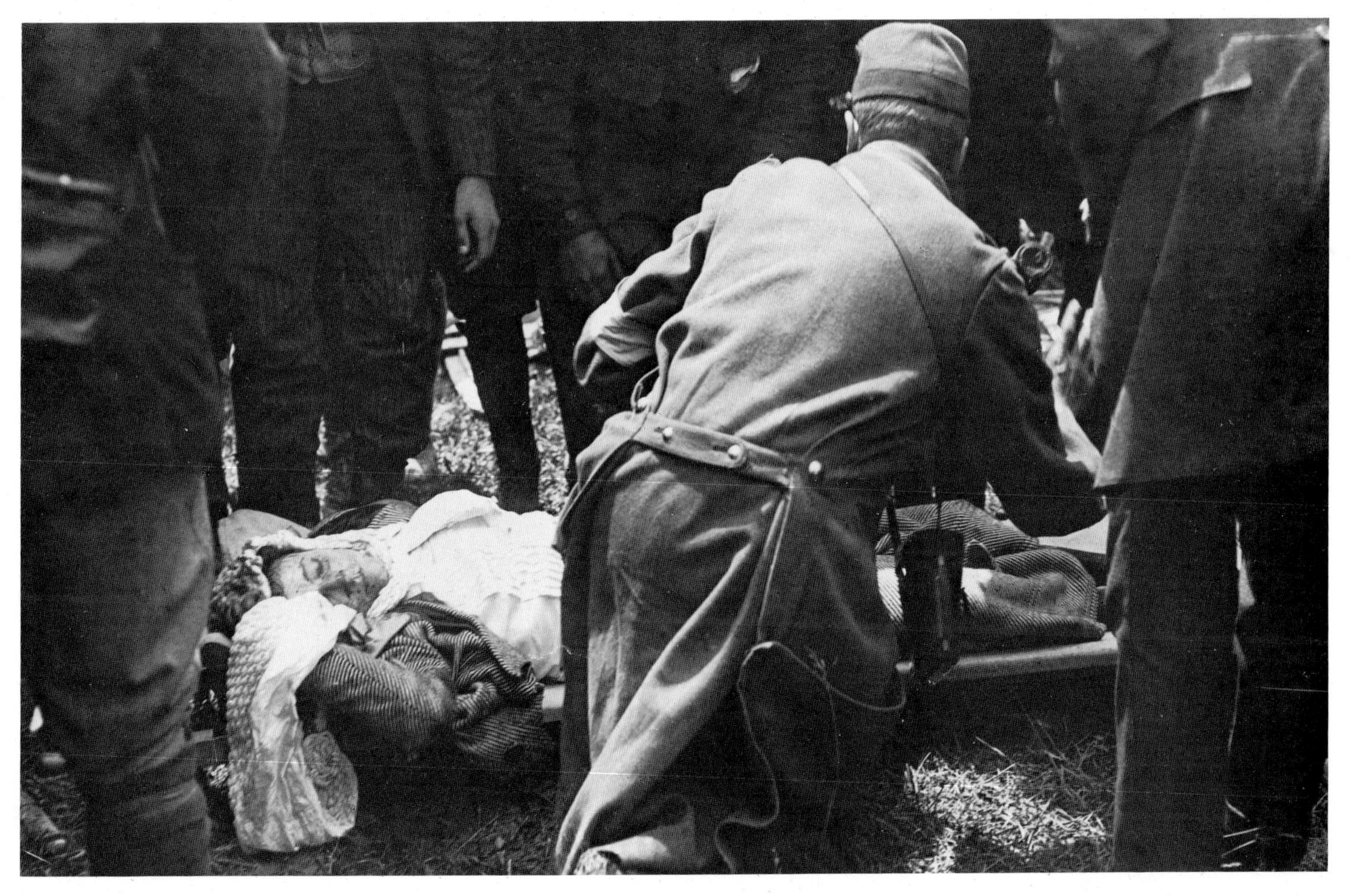

Near death after her crash at Rheims, Raymonde de Laroche is given emergency treatment for multiple fractures. The plucky aviator recovered and was able to resume her racing career.

Almost equally famous was Belgium's first licensed woman pilot, Hélène Dutrieu. Daughter of an Army officer, she began her professional life as a trick bicycle rider. She made her first flight, which lasted 20 minutes, in April 1910, and five months later she astounded Belgium and the world by flying nonstop from Ostend to Bruges, a distance of 28 miles. Capping this feat, she circled a historic belfry spire in Bruges and rose to a height of 1,300 feet, higher than any woman had flown before, while the church bells pealed and thousands of spectators gazed upward from the streets. In May 1911, she competed in a race at Florence, Italy, the only woman in a field of 15 fliers, and outflew all her male rivals to earn the coveted Italian King's Cup. That fall she was in the United States at the Nassau Boulevard Aviation Meet, winning a prize in the women's endurance contest. In December 1911, the daring "Girl Hawk" of aviation—as she was somewhat unattractively called—set a new world nonstop flight record for women by covering 158 miles in 2 hours and 58 minutes. In 1913, she was awarded France's Legion of Honor.

Sadly, the popular press knew her almost as well for the fact that she flew without a corset as for her achievements: Flying without a corset gave her more freedom of movement, she explained, and decreased the chance of injury in an accident. In those early days, all women fliers had difficulty devising costumes that were both ladylike and practical in a cockpit that was exposed to the sky. The corsetless Girl Hawk chose a gray divided skirt, white spats, angora beret and long gauntlets. Her contemporary, Edith Spencer-Kavanaugh of England, favored a form-fitting vest and skirt of such vivid red that she was known as "the Cardinal of the Sky." Eventually, most women fliers decided on loose knee-length golfing trousers, sweaters, high-top boots and soft fabric helmets with goggles.

Blanche Stuart Scott, a small, attractive woman who was born about 1890 in Rochester, New York, was no flying fashion plate—she somehow managed to struggle into three petticoats and covered them with heavy bloomers—but she was the first American woman to make a solo flight. She got into aviation in 1910 by the odd route of driving an Overland car from New York to San Francisco, thus demonstrating on behalf of the Willys-Overland Company that long-distance motoring was so easy even a woman could do it. Impressed by the publicity she got, the manager of the Curtiss exhibition company, founded by pioneer airman Glenn Curtiss, persuaded her to take flying lessons so that she could appear at air shows with the popular Curtiss troupe.

Glenn Curtiss himself had to be talked into teaching her; like Claude Grahame-White, he believed that woman's place was on the ground. Any air accident involving a woman pilot, he thought, would set aviation back for years. Blanche Scott was the first—and the last—woman he agreed to teach.

For three days she listened to theory at the Curtiss Company flying field at Hammondsport, New York, and made short taxiing runs in a 35-horsepower Curtiss pusher—a procedure known as "grass-cutting." She was unable to get up enough speed to leave the ground because Curtiss had blocked the throttle. Then, on September 2, 1910, either a fortuitous wind gust or the connivance of a friendly mechanic gave a boost to the impatient student pilot. "Something happened to the throttle block," she recalled later with a little smile. She rose to a height of about 40 feet, glided briefly and came down to a smooth landing. Within a month she was flying with the Curtiss team—but without benefit of a pilot's license, which was then not required by law.

Billed as the Tomboy of the Air, she flew with various exhibition teams for the next six years. "A thrill every second," promised the handbills. Her most spectacular stunt was her ever-popular Death Dive, featuring a perpendicular plunge from 4,000 to as low as 200 feet. From salary and a percentage of the gate, she made up to $5,000 a week.

Suddenly, in 1916, Blanche Scott retired. "In aviation," she remarked accurately enough, "there seems to be no place for the woman engineer, mechanic or flier. Too often, people paid money to see me risk my neck, more as a freak—a woman freak pilot—than as a skilled flier. No more!"

Because it was never established whether her first brief flight was accidental or intentional, the official honor of being America's first fe-

International gallery of pioneer women

"I shan't be happy till I can fly," declared Hilda Hewlett *(overleaf)* in 1910. She was speaking not only for herself but for women around the globe who yearned to prove they were as qualified as men to pilot airplanes.

Hilda Hewlett won her happiness in August 1911 as the first Englishwoman to earn a flying license. She thus joined a select company of licensed women pilots that already included Raymonde de Laroche of France, Belgium's Hélène Dutrieu, America's Harriet Quimby and Russia's Lidia Zvereva *(overleaf)*.

These women were soon followed by others in Germany *(below)* and Italy *(right)*. In other nations, women were slower to earn their wings, but eventually pioneers in places as distant as China and Brazil became licensed pilots.

Fédération Aéronautique Internationale

ITALIE

(Aero Club d'Italia)

I sottoscritti, rappresentanti del potere sportivo Aeronautico Italiano riconosciuto dalla F.A.I., certificano che: | Nous soussignés, pouvoir sportif reconnu par la F.A.I. pour l'Italie, certifions que:

Mlle Rosina Ferrario

nata a: Milano

il 28 luglio 1888

avendo soddisfatto a tutte le condizioni imposte dalla F.A.I. ha ricevuto il brevetto di: | ayant rempli toutes les conditions imposées par la F.A.I. a été breveté:

pilota AVIATORE (1)

li 3 gennaio 1913

Il Presidente Carlo Montù — Il Segretario Ing. Mina

(1) aviatore, d'aerostato o di dirigibile.

Firma del Titolare

Rosina Ferrario

N.º del Brevetto 203

Tessera 148

Rosina Ferrario appears demure on the first Italian license granted a woman.

Melli Beese's fierce stare reflects her struggle to earn her license in 1911 despite men who twice sabotaged her plane.

Hilda Hewlett possessively holds the 1911 Farman she flew and that was used at the flying school she ran near London.

Brazil's first woman pilot, Anesia Pinheiro Machado, soloed in 1922 at the age of 17 and later set several flying records.

Lidia Zvereva soloed in 1911 and taught other Russian women to fly.

Ruthy Tu became a pilot in China's Army in 1932.

Australia's first licensed woman pilot, Millicent Bryant, coolly coped with extremely rough air on her first flight in 1926.

Bucking Japan's male-dominated society, Tadashi Hyodo worked some two years to earn her license in 1922.

male aviator went not to Blanche Scott but to a remarkable flier named Bessica Raiche. There were not many young women like Bessica Medlar in her hometown of Beloit, Wisconsin, around the turn of the century. She wore bloomers, drove an automobile, excelled in sports, music, languages and the arts, and persuaded her parents to allow her to go to France to study music. She returned home with a French husband, François Raiche, and a head full of the exploits of Raymonde de Laroche and other daring European women fliers.

The Raiches settled in Mineola, New York, and conceived the extravagant idea of building a biplane piecemeal in their living room and assembling it outside. Looking something like the Curtiss pusher that had inspired it, the Raiches' creation was a delicate craft fashioned of bamboo and silk. It was in this machine that Bessica Raiche, without so much as a trial glide or a moment of instruction, rose a few feet off the ground in her first solo flight on September 16, 1910. Within the next several weeks she improved upon her performance and received from the Aeronautical Society a diamond-studded gold medal inscribed to the "First Woman Aviator of America."

The Raiches built and sold two more aircraft and formed the French-American Aeroplane Company—a sort of cottage airplane industry producing planes whose main components were bamboo, piano wire and imported Chinese silk. But Bessica Raiche fell ill, and after her recovery she abandoned aviation to become a practicing physician.

The most celebrated of America's pioneer women fliers was the beautiful and tantalizingly mysterious Harriet Quimby. A willowy, green-eyed brunette, she was reported variously to have been born in Massachusetts, Michigan or California—and into families of either solid wealth or rural poverty. Her public history began in 1902 when she emerged as a dramatic writer for newspapers in San Francisco. A year later she moved to New York as dramatic critic for *Leslie's Weekly*.

It was at the Belmont Park aviation meet of October 1910 that she became seriously interested in flight. An impressive gathering of the world's finest fliers turned up for the international contest, but not one of them was a woman. Harriet Quimby was particularly thrilled by John Moisant's winning race around the Statue of Liberty and asked him to teach her to fly.

When John Moisant was killed a short time later in an air meet in New Orleans, she signed up for flying lessons with his brother Alfred, who had opened the Moisant Aviation School at Hempstead, Long Island. The Moisants' sister Matilde started learning several weeks later, and the two young women quickly became close friends. A naturally gifted pilot, Harriet Quimby applied for and won her pilot's license upon graduation by executing two nearly perfect test flights—each one involving the cutting of figure 8s around two test pylons. She became the first licensed woman pilot in the United States, while Matilde Moisant became the second.

Almost immediately pilot Quimby joined the Moisant International Aviators, described by Matilde Moisant as an organization to advance the science of flying, though it was in fact an exhibition team. Its newest member designed an elegant flying suit of wool-backed plum-colored satin with a monklike hood that became her trademark and was much written about in the newspapers of the time. Wearing this outfit and a pair of rakishly streamlined goggles, she made her professional debut in the fall of 1911 with a moonlit flight over Staten Island, New York, before a crowd of 20,000 spectators.

Newspapers called her the Dresden-China Aviatrix. In fact, she was a very human, totally dedicated flier who worked hard to promote aviation and persuade women to fly. She warned that the United States lagged behind other nations in interest in aeronautic development, and she urged that more attention be given to commercial flight. For *Leslie's Weekly* she wrote a series of articles describing "How a Woman Learns to Fly," "How I Won My Aviator's License" and "The Dangers of Flying and How to Avoid Them." More clearly than most people of her time—men or women—she sensed the potential of aviation as something other than an exhilarating sport. "In my opinion," she wrote, "there is no reason why the aeroplane should not open up a fruitful occupation for women. I see no reason why they cannot realize handsome incomes by carrying passengers between adjacent towns, why they cannot derive incomes from parcel delivery, from taking photographs from above, or from conducting schools for flying." Further, she added, it was entirely feasible that airlines could be established to regularly fly distances of 50 to 60 miles.

During November and December of 1911, still with the Moisant group, Harriet Quimby flew in Mexico City at an aviation meet held in honor of the inauguration of President Francisco Madero. While there, she began thinking about crossing the English Channel. By that time both Louis Blériot and Jacques de Lesseps of France had flown it solo, and her idol, John Moisant, had flown from Paris to London with a passenger in September 1910. In March 1912 she sailed for England and confided her intention to the editor of the London *Daily Mirror,* who offered to underwrite her attempt. She next went to Paris and arranged with Louis Blériot for the loan of a 50-horsepower Blériot monoplane, which she shipped secretly to Dover.

Because of gale winds and rain, she had no opportunity to give her borrowed machine a preliminary test. Nor had she ever used the compass, into the mysteries of which she had been initiated before takeoff by a friend, British aviator Gustav Hamel. He was so skeptical of any woman's chances for a successful Channel flight that he offered to dress up in her plum-colored satin flying suit and make the flight for her: They would meet at some remote landing spot, where he would sneak off and she would sit in the plane and wait for curious French locals to find her. She declined this extraordinary offer, but she listened more attentively to Hamel's parting warning that if she wandered as little as five miles out of the way, she could wind up over the North Sea on a no-return flight that would end in its bitter waters.

There was also the problem, in a plane that offered no protection against the elements, of how to cope with the bone-chilling Channel weather. She solved that, she later revealed in *Leslie's Weekly,* by wearing "two pairs of silk combinations" under her flying suit and over it a long woolen coat, an American raincoat and a sealskin stole. "Even this," she added, "did not satisfy my solicitous friends. At the last minute they handed up a large hot-water bag, which Mr. Hamel insisted on tying to my waist like an enormous locket."

Off at 5:30 on the still and foggy morning of April 16, the Dresden-China Aviatrix climbed steadily in a long circle and reached an altitude of 1,500 feet in 30 seconds. Dover Castle was half-obscured to her view; briefly, she saw the tugboat sent out by the *Mirror.* It was trying to keep ahead of her, but she quickly passed it and plunged into a wall of fog. "I could not see ahead of me at all, nor could I see the water below," she recalled. "There was only one thing for me to do, and that was to keep my eyes fixed on my compass."

Just before setting out from Dover to cross the English Channel in a Blériot monoplane in 1912, Harriet Quimby (right) powders her nose while a companion holds up a mirror for her.

Chilled and soggy, she skimmed through shifting masses of mist at a speed of more than a mile a minute. She dropped from 2,000 to 1,000 feet, trying to find a break in the fog: "The sunlight struck upon my face and my eyes lit upon the white and sandy shores of France." Unable to spot her goal, Calais, she dropped down and landed on the hard and sandy beach. A crowd of fishermen and their families rushed shouting to greet her: "I comprehended sufficient to discover that they knew I had crossed the Channel." She had landed at Hardelot, some 25 miles south of Calais.

Less than three months after her triumph, Harriet Quimby made her last flight. She had arrived at the Harvard-Boston aviation meet at the end of June 1912 with a brand-new, all-white 70-horsepower Blériot monoplane. As part of her performance, she agreed to take meet manager William A. P. Willard—father of Curtiss exhibition flier Charles Willard—out over Dorchester Bay and around the Boston Light.

British aviator Gustav Hamel climbs atop Harriet Quimby's plane to give advice before her Channel flight. Her craft had no brakes, so six men had to hold it while the 50-hp Gnôme engine was running.

Late in the afternoon of July 1, she reassured a solicitous friend that she had no intention of crashing in the bay: "I am a cat and don't like cold water," she said. Then she and her 190-pound passenger took

Hoisted aloft by her friends, Harriet Quimby gets a rousing cheer from the fishermen of Hardelot, France, after her 1912 cross-Channel flight.

off. Blanche Scott was in the air at the time, competing for an endurance prize, and Ruth Law, who had only recently started flying lessons at the Burgess Flying School in Boston, was preparing to take off on her first plane ride.

Some 5,000 spectators watched the Quimby two-seater skimming out over the harbor, flying smooth as a gull in the eight-mile breeze. Easily, she rounded the light, came back over the Squantum airfield at 3,000 feet and circled out over the bay while gradually descending for a landing. Suddenly the Blériot plummeted in a precipitous dive. At "about 1500 feet," recalled Ruth Law many years later, "the passenger just went out of the plane. We were not belted or fastened in any way, at that time. He went up in an arc, out of the plane, as though he had jumped." The pilot fought to get the plane under control but seconds later was thrown out herself. Pilot and passenger hit shallow water within seconds of each other and were crushed to death upon impact with the mud. Blanche Scott, still in the air, witnessed the catastrophe from above and was so shaken that she could barely land. "With an effort plainly visible from the earth," reported the New York *World,* "she turned the nose of her machine downward, came to a landing like a flash, and swooned before anyone could reach her side."

Matilde Moisant's career in aviation was even shorter than Harriet Quimby's had been. Unlike her friend, she thought of flying not so much as a potential career for women but as an amusing sport—even though it had killed her brother John. She attended the Moisant Aviation School for a month, spending a total of 32 minutes in the air. On August 17, 1911, when she became the second woman in the United States to be licensed, she also established a record for the shortest time spent learning to fly—a record that was unlikely ever to be broken. A month later she won a women's altitude competition with a climb to 1,200 feet. Before the end of the year she set a new record of 2,500 feet. Although she was a naturally gifted and extremely daring flier, her career lasted less than a year.

On April 14, 1912, when she was performing acrobatic stunts with the Moisant International Aviators at Wichita Falls, Texas, she came down for a landing to find that spectators had swarmed onto the field. To avoid them, she gave the engine full throttle and soared a few feet over the heads of the crowd but then lost power and crashed. Miraculously unhurt, she was dragged by spectators from the blazing wreckage. The crash, together with her family's concern and her own growing sense of isolation in the male world of aviation, led Matilde Moisant to retire at the age of 26. "My flying career didn't last awfully long," she told an interviewer a few years before her death in 1964, "because in those days that was man's work, and they didn't think a nice girl should be in it."

Given that attitude, it is remarkable not only that so many women flew before World War I but that a few even taught men to fly. In 1912, Germany's Melli Beese opened a flying school in Berlin and taught men

Stretcher-bearers rush to meet a man carrying Harriet Quimby's lifeless body from the water of Dorchester Bay in 1912. "Ambitious to be among the pathfinders," wrote the Boston Post, "she took her chances like a man and died like one."

pilots by the dozens. A woman of great courage, she had scarcely won her license—against stiff male opposition—when she went on to set altitude and endurance records and build her own monoplane, the Melli Beese Dove, which was used as a trainer at her school. Across the Channel, Hilda Hewlett, the first Englishwoman to win a pilot's license, raced against men as early as 1910 and instructed fighter pilots in World War I. In the United States during the same period, Marjorie Stinson became almost legendary as the teacher of Canadian trainees preparing for service with the British Royal Flying Corps.

The Stinsons were a flying family. Encouraged by their mother, two Stinson sons and two daughters became pilots, enabling the family to found the Stinson School of Flying in San Antonio, Texas. Marjorie Stinson earned her license at the age of 20, to become the youngest licensed female pilot in the country. Before she was 22 years old, she had instructed more than 100 Canadian military pilots. A tiny woman,

A flying nurse on a mission of mercy

"I have always believed in the Wings that bring Mercy," wrote Marie Marvingt, a French flier whose long and active career was devoted to humanitarian goals. A nurse by profession, she earned a pilot's license in 1910—one of the first ever granted a woman—and embarked on a lifelong mission to establish an aerial ambulance service for soldiers wounded in battle and for disaster victims.

As early as 1910 she urged the French Army to create a medical aviation branch equipped with special rescue planes carrying nurses, stretchers and surgical supplies. In 1912 she ordered an ambulance plane from the Deperdussin factory in order to demonstrate her ideas, but the company went out of business before the aircraft could be completed. Perhaps because Marie Marvingt was a woman, her proposal was for years largely ignored by the French military. "What skepticism didn't I encounter," she later recalled, but "no criticism, no chaff was able to change my vision."

During the First World War a few French Army units used planes for emergency medical service, although not on a regular basis. Marie Marvingt continued her campaign after the Armistice. In the early 1930s she helped set up classes to train nurses and pilots for aerial medical work, and at last, in 1934, the Army invited her to organize a civilian air ambulance service in the French protectorate of Morocco. Soon airborne nurses there were busily engaged in bringing to fruition the mission of winged mercy that she had first envisioned a quarter of a century earlier.

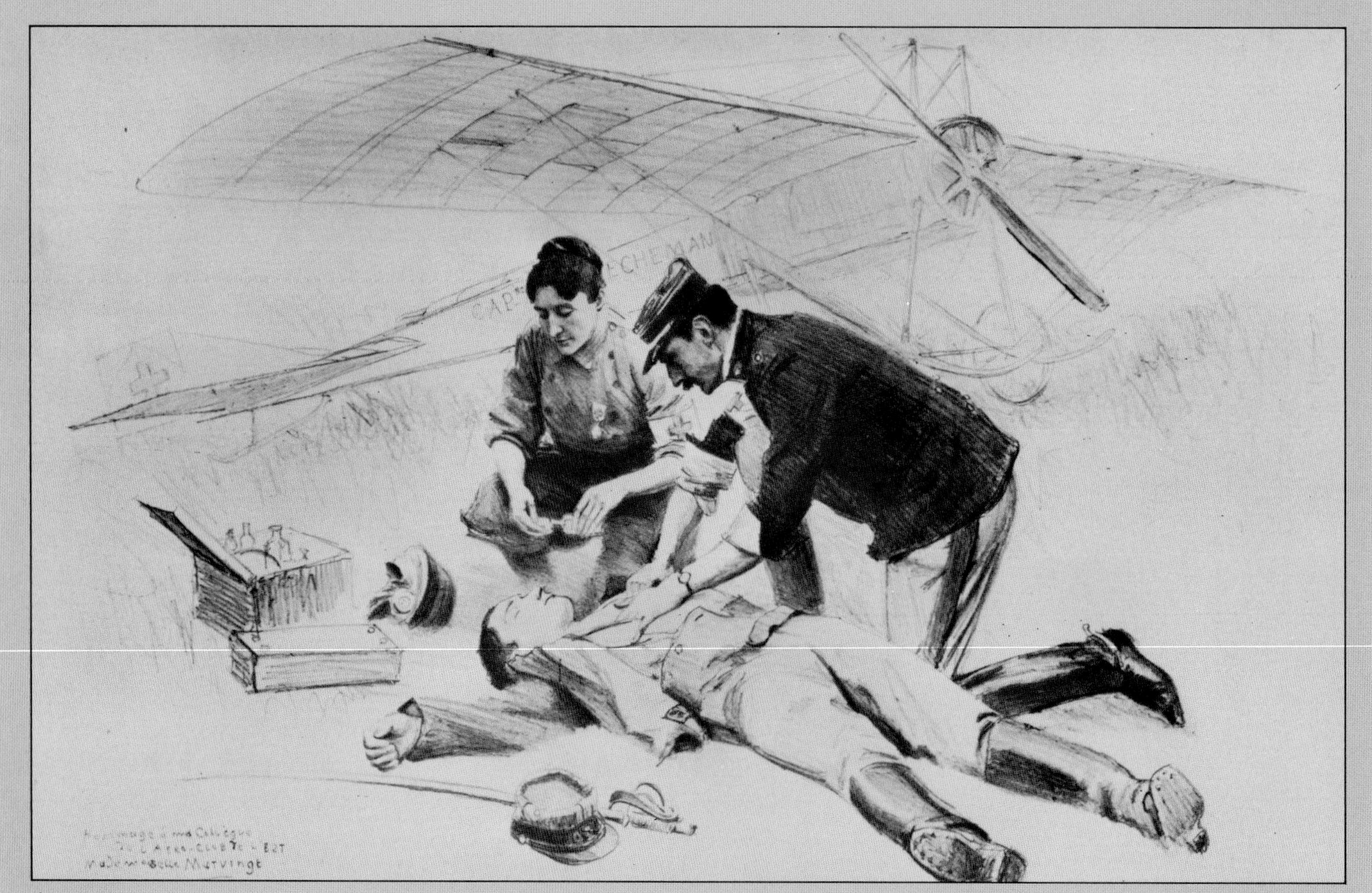

A nurse-pilot and a medic aid a wounded soldier in this 1914 drawing depicting Marie Marvingt's plan for an aerial ambulance corps.

she made what she called "a gentleman's agreement" with her students that they would instantly release the controls whenever she told them she wanted to take over, thus avoiding struggles that she would almost certainly have lost and that could easily have resulted in the loss of the plane as well.

Marjorie's sister, Katherine, contributed to the family school not by teaching but by stunt-flying to raise money for its support. As the fourth licensed woman pilot in the United States, Katherine Stinson attracted attention early and soon was known as the most daring female stunt flier of her generation. In 1913, she became the first woman to fly the mail—from the fairgrounds outside Helena, Montana, into the city proper. In succeeding years she set successively longer endurance and distance records and toured England, China and Japan, giving exhibitions to admiring crowds *(pages 34-41)*.

Perhaps her most famous flight was from San Diego to San Francisco in 1917. "Passing over Los Angeles, I began to rise gradually over Tehachapi Pass, 9,000 feet above," she recalled. "I knew that aviators had tried to cross it and failed and I knew, too, that once over the top I would have no trouble." She cleared it and indeed had no trouble the rest of the way—cruising along at 62 miles per hour, waving at trains and school children and finally coming down to be greeted by the cheers of soldiers assembled at the Presidio Army base in San Francisco. She had flown 610 miles, a new nonstop distance record for both men and women. "I'll bet Ruth Law is glad a girl and not a man broke her record," she said.

She was referring to a flight that Ruth Law had made a year earlier to extraordinary national acclaim. No woman flier of her time attracted more attention than Ruth Law or served as a greater spur to the ambition of other fliers. She soloed one month after seeing Harriet Quimby plummet to her death. She then set out on a record-breaking career with the encouragement of two men: her brother, Rodman Law, who was known as the "Human Fly" for his work as a trick parachutist and stunt man for the Pathé film and newsreel company, and her husband, Charles Oliver, a devoted fan and a superb business manager.

She received her license in November 1912. An intensely competitive woman, she needed only to be warned against doing something to persuade her to try it. In 1915 she gave her first exhibition of acrobatic flying at Daytona Beach, Florida, and announced that she was going to loop the loop for the first time. Charles Oliver was, as usual, watching her and worrying. She recalled many years later that he said: "Now, Ruth, you don't have to try to loop, if you don't want to." Given her stubborn temperament, she said, "that was enough to make me figure I was going to do it. So the next morning I went out on the beach and I got up to about 1,500 feet. I dipped the plane back, and pulled back the elevator in what seemed to be a nice circle, straight up over, and the plane went over like a bird. I had to laugh, it was so easy. So then I made another loop just to be sure that I could, and I came down to great excitement on the beach. My husband said, 'You didn't have to do it the *second* time!' "

At an aviation meet at Sheepshead Bay, New York, in the late spring of 1916, she entered two altitude contests. Victor Carlstrom won the first by climbing to 14,500 feet. Ruth Law took second place with 11,000 feet. On her next try she became so chilly in the upper air that she could no longer handle the controls efficiently, but when she came down she was certain she had topped her previous altitude. She had, by 200 feet. But this time she was second best to Steve MacGordon, who had reached 15,800 feet. She was furious: Her aim, then as always, was to set a record that would stand not just against women but also against men. That competitive spirit soon led her to her greatest flight.

On November 2, 1916, Victor Carlstrom had been forced down at Erie, Pennsylvania, while attempting a nonstop flight from Chicago to New York. Although he had not reached his goal, he had nevertheless set an American cross-country nonstop record of 452 miles. Ruth Law immediately decided she could better it. Using auxiliary tanks on her aging 100-horsepower Curtiss pusher, she upped the gas capacity from eight to 53 gallons and installed a rubber fuel line that presumably could not break, as Victor Carlstrom's had. Since her plane was open to the elements, she had her mechanics install an aluminum windshield to protect her feet and legs from the wind. For instruments she would be using standard equipment such as a clock, a compass and a barograph, to which she added a novel navigational device—a rolling-map-display case, containing a loop of maps covering her intended route.

She had decided that her Geodetic Survey maps were too big for her to handle in the open air. "So I cut strips out, about eight inches wide," she reminisced, "and pasted them on a strip of thin cloth." Then she had a camera-like box made with two rollers and a glass top, and wound the map strip onto the rollers. With this contraption tied to her left knee she could turn the map-case knob with her right hand to keep her route constantly in view. She also provided herself with a backup navigational system: "I had compass markings on the cuff or gauntlet of my gloves, from one city to the next."

Thus, flying partly out of an ingenious map case and partly off the cuff, Ruth Law left Chicago at 8:25 on the morning of November 19—within three weeks of Carlstrom's flight. She flew on through blustery weather until her engine began to sputter. Nearly six hours and 590 miles after takeoff, she landed on a field at Hornell, New York. She was short of her New York City goal, but she had set two new records: the American nonstop cross-country record for both men and women, and the world nonstop cross-country record for women.

At Hornell she picked up gas and had her plane's spark plugs changed by a young Army lieutenant named Henry H. "Hap" Arnold; he would one day become wartime commander of the United States

This "bombshell" leaflet, which bears Ruth Law's picture on one side and a blunt message on the other, was one of many that the famed woman flier dropped from the air during a 1917 Liberty Loan drive in the Midwest. But the Army declined her offer to serve as a combat pilot.

Army Air Forces. Then she flew eastward to Binghampton where she spent the night. Early next morning she took off for New York City, planning to land on Governors Island in Upper New York Bay.

Crowds of officials and photographers had gathered by the time she skimmed over the tall buildings of Manhattan, heading downtown toward the southern tip of the city. Again, her engine began to sputter ominously, but now there was no vacant field for her to land on. Coolly, she rocked her plane up and down to tilt the remaining fuel from the tank into the carburetor; and with the last drops of gas she pulled back on the controls to lift the little Curtiss pusher as high as she could before it began to fall. Then the engine cut out altogether. Completely out of fuel, Ruth Law made a long, sweeping glide down to a safe landing on Governors Island.

The City of New York acclaimed her for weeks. President and Mrs. Woodrow Wilson attended a dinner in her honor on December 2; and on December 18 the Aero Club of America and the New York Civic Forum paid tribute to her with a dinner at the Hotel Astor attended by such notables as Admiral Robert E. Peary and Captain Roald Amundsen, discoverers of the North and South Poles, and Corinne Roosevelt Robinson, sister of former President Theodore Roosevelt. "Miss Law's splendid accomplishment," said Admiral Peary, "has shone so that the whole world may read what a woman can do." More to the point was a speech by a popular novelist named Eleanor Gates. "It is easy," she said, "to get a dinner if you are a man. You get one if you are a such-and-such degree Mason, or a naughty Elk, or just because it's time to have another dinner. But for a woman to sit in glory at the Hotel Astor, she must do something superhuman."

When the United States entered World War I in 1917, Ruth Law wanted to fly for the United States Army but was turned down: "We don't want women in the Army," said Newton Diehl Baker, Secretary of War. The experience of women in other countries was much the same. The two who came closest to combat duty were Hélène Dutrieu in France and Princess Eugenie Mikhailovna Shakhovskaya in Russia, both of whom served as reconnaissance pilots. In the United States, Ruth Law and other top women pilots, such as Katherine Stinson, were restricted to recruiting tours and fund-raising flights.

Stung by her peremptory rejection, Ruth Law in 1918 wrote an article titled "Let Women Fly!" for the magazine *Air Travel.* "It would seem," she noted, "that a woman's success in any particular line would prove her fitness for that work." But that was obviously not the case: "There is the world-old controversy that crops up again whenever women attempt to enter a new field—Is woman fitted for this or that work?"

By the beginning of World War I, women in aviation had repeatedly answered the question in the affirmative by their accomplishments. Yet the question persisted. It would continue to haunt the next generation of women fliers, discouraging many but goading others to achievements that tested the limits of women's role in the air.

Winged idol in the East

In 1916, at 25 years of age, flier Katherine Stinson was, physically, a mere wisp of a woman. ("I weigh only about 101 pounds," she was fond of noting. "I'm very particular about that one pound.") But during an exhibition tour of Japan and China that she began late that year, her apparent frailty only increased her impact on Oriental audiences. Her first appearance drew 25,000 people to Tokyo's Aoyama Parade Ground. They were thrilled by the fearless pilot's breathtaking aerobatics, but the real sensation, in a society whose customs severely restricted the activities of women, was her sex.

"The women have simply overwhelmed me with attention and seem to regard me as their emancipator," wrote the American pilot. She described the mob scene around her plane after the first Tokyo flight: "The women were wild with enthusiasm and the men were not far behind." Hastily organized female fan clubs deluged her with gifts and enough flowers to fill two rooms of her hotel suite.

She also elicited the sincerest form of flattery—imitation. After watching Katherine Stinson fly loops over Tokyo, a young woman named Komatsu Imai learned to fly and managed a seven-year flying career. Would-be women pilots flooded one Tokyo flying school with applications, but only one applicant was accepted, and she failed to get her license.

When Katherine Stinson returned to the United States in May of 1917, she left behind her not only a heightened interest in flight among the Chinese and Japanese people but a dawning awareness among their women of her often-expressed philosophy: If other people can fly, she said, "I don't see why I can't."

Youthful Chinese admirers, part of a crowd that includes some turbaned Indians, doff their caps to an appreciative Katherine Stinson as she climbs from her cockpit in Peking in March 1917. The press of well-wishers was typical of her many appearances in China and Japan.

At the controls of her Laird Looper, rented from plane builder E. M. Laird, Katherine Stinson takes off from Tokyo's Aoyama Parade Ground.

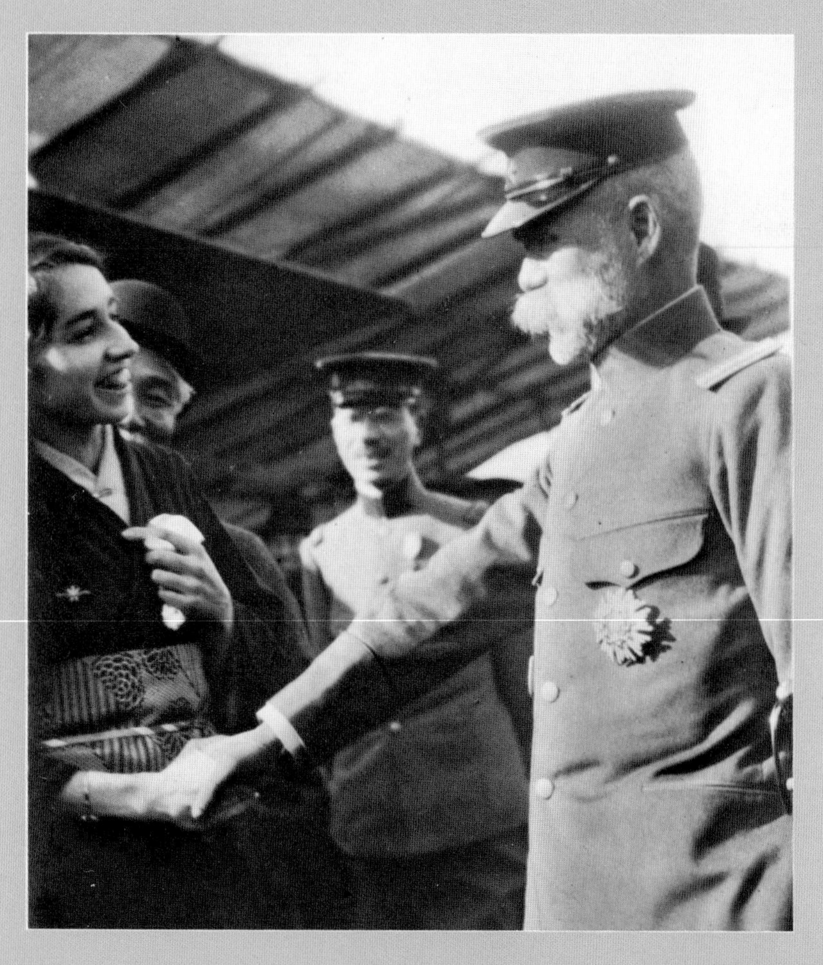

A driving force behind the accelerating development of Japanese Army aviation, Lieutenant General Gaishi Nagaoka extends his hand to the visiting flier. Army aviators, impressed by her fearless aerobatics, were keenly interested in the abilities of both the pilot and her aircraft.

Watched by the ever-present Japanese, the demurely dressed flier leans against the propeller of her biplane. She was careful about her appearance after a flight, hoping to look "as if I hadn't lost my feminine qualities, even though I am an aviator."

An Osaka geisha presents a floral tribute to the adventurous flier from another world.

Katherine Stinson waves gaily from a throng of the curious in Tokyo. A newspaper in her hometown of San Antonio, Texas, reported: "Representatives of the press have accompanied her almost everywhere; and the population in general have shouted 'banzai' at her approach."

Concerned bystanders help an uninjured Katherine Stinson from her plane after she overshot a runway in Shanghai in early 1917.

The American flier banks over Peking's Temple of Agriculture as spectators watch from the vantage point of a bridge. During the flight the rudder control snapped off; she had to bend down to the floor and manipulate the stub to change direction, then pop up to see where she was going.

2

Challenging a deadly ocean

For several days in the late summer of 1927, the world focused its attention on one daring woman and a heavily laden plane. The woman was 63-year-old Princess Anne Lowenstein-Wertheim—a prominent member of the British peerage who had married Prince Ludwig Karl of the south German principality of Lowenstein-Wertheim-Freudenberg. The plane was a single-engined Fokker monoplane named the *St. Raphael.* Early on the morning of August 31, the Princess had arrived at Upavon Airdrome on the Salisbury Plain and had told the assembled press that she was proud "to be the first woman to attempt the Atlantic crossing." She then kissed the episcopal ring of the Archbishop of Cardiff, climbed into the makeshift wicker passenger seat of the *St. Raphael* and took off for Ottawa with a hired pilot and copilot in the cockpit. Her plane was sighted once over the Atlantic and never seen again.

What was remarkable about the flight was not its tragic end—the Atlantic swallowed no fewer than 19 fliers in the year 1927 alone—but the fact that the first woman to make the attempt should have done so as a passenger. The Princess was herself a licensed pilot with some experience in long-distance flights, but it apparently occurred neither to her nor to the pilots she hired that a woman flier might possess the skills to challenge the mighty Atlantic. To many, her disappearance merely confirmed that women had no place in the great exploratory adventure that aviation was becoming.

World War I had given flight a new technology. Planes were larger, faster, sturdier and more reliable than the insect-like contraptions of aviation's first decade. It was becoming possible, if neither safe nor easy, to fly great distances, from country to country and even from continent to continent. To a generation of fliers emerging from the War, the supreme challenge was the North Atlantic—an immensity that stood as both barrier and link between Europe and the New World. The first to conquer it in a nonstop flight were England's Captain John Alcock and Lieutenant Arthur Whitten Brown, who crossed from Newfoundland to Ireland in June of 1919. Other aviators tried over the next several years without success.

If women were rare among early long-distance fliers, the reasons were both social and financial. The military flight training that prepared so many of the early transcontinental fliers was closed to women—as were most flying jobs and access to the kind of financial backing that

Amelia Earhart wears a leather helmet for her photograph on the flying license issued her in 1923 by the prestigious Fédération Aéronautique Internationale. Although an FAI license was not legally required, she sought it as recognition of her flying ability—an ability that the whole world would recognize within a decade, as she became one of the best-known aviators in history.

made long-distance flying feasible. Then, too, women doubted their own capability to undertake long flights. The rare exceptions to that rule were women of extraordinary character.

One was a brash and cheerful young Frenchwoman named Adrienne Bolland. Inspired by tales of France's pioneering women fliers, she took lessons at a flying school run by the Caudron aircraft-manufacturing company and got her license in 1920. She became an accomplished stunt pilot (a few years later she would set a record by performing 212 consecutive loops) but soon began looking for another way to test her new skills. When she learned that no woman had yet flown over the Andes, she arranged for the Caudron company to ship one of its biplanes to South America so that she could fly it through the mist-enshrouded peaks that separate Argentina from Chile. Although she was forewarned that her plane had a ceiling of 13,000 feet and that the lowest peak in the Andes was 14,000 feet, she had no idea, she later recalled, "of exactly what I was in for." She took off from Mendoza, Argentina, early on April 1, 1921. Groping her way through the mountain passes and almost losing consciousness from the altitude and extreme cold, she completed the exhausting trip after 10 grueling hours of flying.

It was seven years before other women undertook flights of comparable difficulty. In 1928, two titled British women astonished the aviation world by setting records up and down the African continent, flying over country where planes had rarely been seen. One of them, Lady Mary Bailey, daughter of an Irish peer, became the first woman to fly solo from England to South Africa; in a de Havilland Moth, she followed a route that took her across the Sudan and down the edge of Lake Victoria to Tanganyika (part of present-day Tanzania) and on to Cape Town. She then turned around and flew back over the Belgian Congo (now Zaire) and the Sahara. At Khartoum, she crossed paths with Lady Heath, the former Sophie Eliot-Lynn, who, in an Avro Avian monoplane, was making the first solo flight from the Cape of Good Hope to Cairo. Both women went on to complete their flights and afterward made light of their achievements. Lady Heath—who went equipped with a Bible, a shotgun, tennis rackets, six tea gowns and a fur coat—declared that "it is so safe that a woman can fly across Africa wearing a Parisian frock and keeping her nose powdered all the way."

In the United States, the contemporaries of these remarkable women were largely restricted to stunting and exhibition flights that tested their nerve but not their potential. Two of the best of them—Laura Bromwell of Ohio and Bessie Coleman of Texas, the first licensed black woman pilot in the world—were killed in stunting crashes at the height of their careers. Other women, like Ruth Law, abandoned flying, both because of the risks involved and because it was becoming apparent as airline services developed that there was little place in commercial flying for women. When women on both sides of the Atlantic began planning transatlantic flights after Charles Lindbergh's spectacular crossing of

A quest for equality in the air

"If I can create the minimum of my desires," Bessie Coleman told her family, "there shall be no regrets." But for Bessie Coleman, a black woman who wanted to fly airplanes, even the minimum goals seemed worlds away behind barriers of racial and sexual prejudice.

Born into a poor Texas family in 1893, she did well in school. When poverty kept her from finishing college, she drifted to Chicago, where she opened a chili parlor. Though the business made money, she was discontented, and after seeing barnstorming stunt pilots, decided to learn to fly. Rejected by American flight schools because of her race, she used her savings to go to France and there earned her pilot's license.

She returned to America in 1921, eager to open a flight school and "give a little coloring" to aviation. She believed that "the air is the only place free from prejudices" and yearned to introduce other blacks to that special environment. She was denied her dream, however, when she died in a crash in Florida in 1926 while rehearsing for an air show.

Wearing a pseudo-military flight suit, Bessie Coleman stands before a Nieuport, the kind of aircraft the pioneer black aviator learned to fly in France.

May 1927, they cast themselves not as pilots but as patrons and as passengers going along for the ride.

Those who knew her were not surprised that Princess Lowenstein-Wertheim aspired to be the first of these patron-passengers to cross the Atlantic by air. A sportswoman, inventor and iconoclast, she had for several years displayed a devotion to aviation that alarmed her family and friends. When she had the *St. Raphael* built and declared her intention of crossing the Atlantic, they tried without success to stop her. Her subsequent disappearance provoked a public outcry about "stunt" and "suicide" missions but did not discourage other women from making similar attempts. Two of the most celebrated were a handsome 23-year-old woman from Alabama named Ruth Elder and a well-to-do New York divorcée in her mid-thirties named Frances Wilson Grayson.

Ruth Elder was a competent flier who eventually would demonstrate her skill by placing fifth in the first Women's Air Derby, in 1929, before going on to Hollywood and a screen career. But in August of 1927, when she announced plans to fly the Atlantic, she was only a student pilot: Most of the flying, she admitted, would be done by her flight instructor, George W. Haldeman. Although the flight was an obvious publicity stunt (the two were promoting a Stinson Detroiter plane they had christened *American Girl),* it received enormous newspaper coverage that stirred the ire of women's groups angered by such a blatant attempt to cash in on a pretty face. The flight also touched off a vogue for "Ruth ribbons" of the kind Ruth Elder used to tie her hair.

Frances Grayson's motives were perhaps purer: She wanted to prove that if men could fly the Atlantic so could women. But she, too, was forced to admit reluctantly that her own flying skills were not up to the task and that she had hired an all-male crew. The two women kept their respective planes at Curtiss Field on Long Island, and during the fall of 1927 they anxiously watched each other's test flights while waiting for a break in the weather. Frances Grayson, who cherished an unfounded belief that the planes of the late 1920s were reliable enough to fly the Atlantic at any time of the year, was determined to make the crossing by the shorter, northern route; Haldeman favored the southern route via the Azores to avoid the danger of icing.

George Haldeman and Ruth Elder got off for Europe first, on October 11, and seemed on their way to succeeding when an overheated engine forced them to ditch northeast of the Azores near a freighter, which rescued them. When Frances Grayson heard the news, she was undeterred: "The theory about there being a season for transatlantic flying is wrong and I will prove it wrong," she said. "I am going to be the first woman to fly across the Atlantic and mine will be the only ship since Lindbergh's to reach its destination. I will prove that woman can compete with man in his own undertakings."

Having flown to Maine, she and her crew tried three times to take off in their Sikorsky amphibian, *Dawn,* from Old Orchard Beach; three times they were driven back to the beach by technical difficulties and foul weather. With fall deepening into winter, they decided to shift operations to Harbour Grace, Newfoundland, which was 900 miles closer to Europe. First they had to go back to New York to have the plane overhauled. From there the *Dawn* took off for Newfoundland on December 23. It never arrived and never sent any clearly identified messages. Presumably, Frances Grayson and her crew perished in the tempestuous Atlantic.

Remarkably, there was no dearth of transatlantic candidates to replace those who disappeared. One of them, an American socialite named Mabel "Mibs" Boll, spent an entire year trying futilely to find a pilot who would fly her across the Atlantic for $25,000, settling at last for a flight to Havana instead. In England, the Honorable Elsie Mackay, actress, horsewoman, pilot and daughter of shipping magnate Lord Inchcape, bought a single-engined Stinson Detroiter and hired a distinguished wartime and commercial pilot, Captain W. G. R. Hinchliffe, to help her fly it across the Atlantic from east to west. After taking off from England in March 1928, they were never heard from again.

Yet out of this record of seeming futility, the most illustrious career in women's aviation was born. The instrument of its birth was the wealthy Mrs. Frederick E. Guest of London—formerly Amy Phipps of Pittsburgh—who in 1928 bought a three-engined Fokker from Commander Richard E. Byrd, renamed it *Friendship* and hired a pilot to fly her across the Atlantic. When her family strenuously objected, she agreed reluctantly to withdraw—on the condition that "an American girl of the right image" be found to take her place. On the committee to find this paragon was George Palmer Putnam, publisher of such renowned aviation books as Lindbergh's *We* and Byrd's *Skyward*. One of Putnam's associates asked retired Rear Admiral Reginald K. Belknap of Boston if he had any ideas for candidates. "Why, yes, I know a young social worker who flies," said the admiral. "Call Denison House and ask for Amelia Earhart."

Ruth Elder and George Haldeman test their newly invented buoyant rubber safety suits before attempting a transatlantic flight in 1927. Although a ruptured oil line forced the fliers' single-engined Stinson into the sea, a Dutch tanker rescued them before use of the suits became necessary.

Putnam and his associates could not have found a better candidate. Daughter of a Kansas railroad lawyer whose career was ruined by alcoholism, Amelia Earhart had spent her early years in Kansas City. By the time she was 11, the family had resettled in Des Moines, Iowa—the first of a series of moves that grew more frequent as the family fortunes declined. "There are two kinds of stones, as everyone knows, one of which rolls," she recalled. "Because I selected a father who was a railroad man it has been my fortune to roll."

The period beginning in Des Moines, Amelia's younger sister, Muriel, wrote long afterward, "saw the loss of our material prosperity and the beginning of the disintegration of the family." The problems at home, she added, "made an indelible impression upon us and help to explain some of Amelia's actions and attitudes in her later life." Certainly the elder girl would grow up to be an independent, striving young woman,

Alongside the American Girl, the plane she hoped would carry her to Paris, Ruth Elder hefts the picnic basket she packed for the trip. She said she had two reasons for going: her wish to become the first woman to accomplish the flight and her desire to buy an evening gown in Paris.

with a lifelong aversion to alcohol and a compulsion to break with the patterns of the past.

In 1916 Amelia Earhart graduated from high school in Chicago and prepared to enter Bryn Mawr. But her plans for college abruptly changed when she visited her sister in Toronto during the winter of 1917-1918. Shocked by the sight of World War I casualties hobbling along the streets, she took a concentrated course in first aid with the Canadian Red Cross and spent the rest of the War as a nursing aide in the Spadina Military Hospital in Toronto. She worked in the dispensary, made up patients' beds, carried trays of food and emerged from the experience a confirmed pacifist.

One day she made a visit to the local flying field to watch military aircraft take off. "It was November, with snow on the ground," she recalled. "The plane, on skis, started. The backwash of the propellers threw burning, biting snow into my face. It was like the finest cold shower you ever imagined. I determined then that I would some day ride one of these devil machines."

In the fall of 1919 she enrolled as a premedical student at Columbia University in New York. Convinced after a year that she did not want a career in medicine, she moved west to join her parents in Los Angeles—which was becoming the center of the nation's nascent aircraft industry. Soon she was working at a succession of odd jobs to pay for flying lessons. "The family scarcely saw me," she recalled. "I worked all week and spent what I had of Saturday and Sunday at the airport a few miles from town."

She got her license and appeared in a few local air meets—in one, in 1922, she set a new women's altitude record of 14,000 feet. But she "didn't like public flying" and still had no idea what to do with her aviation skills. When her parents' marriage broke up in 1924, she moved east to Boston and drifted for two years between college courses and teaching jobs. She was three months short of 31, and engaged in social work with immigrants at the Denison settlement house, when she received the astounding invitation to fly the Atlantic as a surrogate passenger for Mrs. Frederick Guest.

Amelia Earhart understood from the first that she would have little hand in actually flying the plane; in the cockpit would be two veteran pilots—Wilmer Stultz and Louis "Slim" Gordon. Although some of her admirers have criticized her willingness to fly as a passenger, the fact is that she lacked the experience of instrument flying and of multiengined planes that the *Friendship's* backers considered to be essential to transatlantic flight. Realizing this, she was grateful for the opportunity to cross the Atlantic under any conditions. "When a great adventure's offered, you don't refuse it," she said.

She signed on in April of 1928 and went to look at the plane that would carry her over the Atlantic: "When I first saw *Friendship* she was jacked up in the shadows of a hangar in East Boston. Mechanics and welders worked nearby on the struts for the pontoons that were shortly to replace the wheels. The ship's golden wings, with their spread of seventy-two feet, were strong and exquisitely fashioned." The orange color of the fuselage, which, she noted, "blended with the gold," was chosen not for its beauty but for maximum visibility in case of a forced landing at sea.

The *Friendship* was ready to go by mid-May; for three weeks the fliers awaited a break in the weather. The passenger-designate went on with her work at Denison House and stayed away from the airfield, afraid that her presence around a plane being fitted for oceanic flight might lead to publicity and a transatlantic race. So strict was the secrecy surrounding the flight that she did not even let her parents know she was leaving. Instead, she wrote them what she wryly called "popping-off letters"—farewell notes reflective of her philosophy, to be delivered only in the event of a fatal crash. To her mother, she expressed her belief in the importance of living with, even embracing, a measure of risk: "Our family tends to be too secure. My life has really been very happy, and I don't mind contemplating its end in the midst of it." With her

The early years of an airborne tomboy

In Kansas at the turn of the century, Edwin and Amy Earhart doted on their tomboy daughters, Amelia and Muriel. Edwin gave the girls footballs and rifles, while Amy shocked the community by dressing them in gym suits instead of skirts. Edwin's job caused the family to move from town to town, and the girls' interest in rough sports and shooting rats raised eyebrows wherever they went.

Amelia's parents did not pressure her to reform as she grew older, even when she dabbled in the male domains of science and automobile mechanics. But in 1920, when she went aloft at an air show and returned home hell-bent on learning to fly, even her liberal parents hesitated. They soon gave in, however, and within months Amelia was flying a Kinner Airster her mother helped her buy.

Amelia was born in this house, her grandparents' home in Atchison, Kansas, in 1897.

Six-year-old Amelia (right foreground) celebrates the Fourth of July at a party that reflects the social climate of Kansas in 1904.

Amelia (center) and Muriel (left foreground) visit their mother's family in Atchison during a vacation with their parents (right).

Enjoying a perquisite of his job as a railroad executive, Edwin Earhart takes Amelia, Muriel and the family cook (right) to Atchison in the company railroad car that was reserved for his use.

A volunteer nurse's aide during the First World War, Amelia Earhart wears a nursing smock to serve wounded soldiers in a Toronto military hospital.

Amelia (third from right) learned auto mechanics in this college class.

In order to earn the fee for flying lessons, Amelia Earhart dons businesswoman's clothing to take a job with the telephone company.

The two sisters bask in success on a Santa Monica, California, beach in 1923, Amelia (right) having won her pilot's license and Muriel her teaching degree.

Amelia Earhart joins her flying instructor, Neta Snook (left), in front of a Kinner Airster biplane that the latter taught her to fly. The two women became friends, flew together frequently and double-dated.

father, she was jaunty: "Hooray for the last grand adventure! I have no faith we'll meet anywhere again, but I wish we might. Affectionately, your doter Mill."

Before the *Friendship* was finally cleared for takeoff on June 3, the woman crew member borrowed a fur-lined flying suit from a friend without explaining what she wanted it for. In the gray predawn the *Friendship* taxied across Boston harbor with Stultz at the controls. "The dream became actuality," she later wrote. "We were off!"

The plan was to fly to Trepassey, Newfoundland, refuel and jump off from there on the longer, transatlantic leg of the trip. Bad weather first forced them down at Halifax and later bottled them up in Trepassey for nearly two weeks. George Putnam, who had remained active in preparations for the flight, wired her suggesting YOU TURN IN AND HAVE YOUR LAUNDERING DONE. That was unnecessary, she wired back: SOCKS UNDERWEAR WORN OUT. SHIRT LOST TO SLIM AT RUMMY. She signed herself "AE"—which is what she preferred to be called for the rest of her flying career.

After the *Friendship* at last lifted off from Trepassey on June 17, 1928, the flight itself was almost an anticlimax. For hours, the trimotor throbbed on through a wet, gray world, at one point flying so low under a 500-foot ceiling that the spray from the turbulent sea nearly splashed the pontoons. AE kept the log, crouched in a little compartment behind the extra fuel tanks. Occasionally she changed places with Gordon for an up-front view of the drifting fog and an aerial beauty she had not seen before. With typical self-mockery, she noted that she was "getting housemaid's knee kneeling here gulping beauty."

By dawn, the weary Stultz had been flying for 19 hours, his radio was dead and he was not sure where he was. When he estimated that he had about an hour's fuel left, he sighted an ocean liner and circled while AE dropped a message in a bag weighted with an orange asking that bearings be painted on the ship's deck. The bag missed; the *Friendship* flew on.

Half an hour later Stultz sighted fishing vessels and then a coastline and dropped smoothly into a sheltering bay: To his astonishment, he found that in the overcast the *Friendship* had flown clear over Ireland and set down at Burry Port, Wales. The plane had been in the air for 20 hours and 40 minutes.

Its arrival made Amelia Earhart a celebrity. It was not a role that she cherished, for she knew exactly what her contribution to the trip had been. "All I did," she said dryly, "was lie on my tummy and take pictures of the clouds." Nonetheless, she was regarded as the star of the flight, while the men in the cockpit were practically ignored. President Calvin Coolidge sent her a cable of congratulations, to which she wired back: SUCCESS ENTIRELY DUE GREAT SKILL OF MR. STULTZ.

But it was a losing battle: "From the beginning," she wrote, "it was evident the accident of sex made me the chief performer in our particular sideshow." The first two weeks after the landing, she recalled, were

After becoming the first woman to cross the Atlantic Ocean by air, Amelia Earhart peeks from the Friendship (top) as the seaplane is tugged across Southampton harbor. At left, the fliers are boated ashore with their sponsor, Mrs. Frederick Guest (left). Above, a souvenir badge celebrates the three aviators.

"a jumble of teas, theaters, speech making, exhibition tennis, polo and Parliament, with hundreds of faces crowded in." Back in the United States, the fliers were invited to make official visits to 32 cities—with the crew's only woman member again as the star. She was inundated with speaking invitations and offers of employment.

In fact, the *Friendship's* sponsors had chosen even more shrewdly than they knew when they yanked Amelia Earhart out of social work and thrust her into the mainstream of aviation. Bright and articulate but becomingly modest, she possessed the kind of mysterious magnetism that so often is an ingredient in the making of public idols. The press was quick to note her resemblance to Lindbergh—to her dismay, they dubbed her Lady Lindy—yet her look and style were distinctly her own. Although she thought she had a face "that looked like everybody's," it was not long before her famous smile and mop of hair were instantly recognizable. She introduced a new look with her tailored slacks, her simple dresses with a single strand of pearls, and above all, her tousled short hair—a trademark that over the years became so famous it prompted famed Kansas editor William Allen White to plead in an editorial: "Comb your head, kid, comb your head!"

She was famous—and far from satisfied. On the verge of a career that never would have been possible except for the lucky accident of the *Friendship* flight, she felt that she still had to prove her worth as something more than mere baggage. Sadly, she acknowledged the truth of editorial comments, like one published in the London *Evening Standard,* that her presence on the *Friendship* crossing had added no more to its success than the presence of a sheep might have done.

Yet her flight had accomplished something intangible that she herself only gradually became aware of. It not only galvanized women's interest in aviation but gave women aviators a mark to surpass. If a woman

Ten contestants in the 1929 Women's Air Derby gather behind trophies for which they would compete. They are (from left) Louise Thaden, Bobbi Trout, Patty Willis, Marvel Crosson, Blanche Noyes, Vera Walker, Amelia Earhart, Marjorie Crawford, Ruth Elder and Florence "Pancho" Barnes.

had flown the Atlantic as a passenger, the next time she must fly it alone. In Amelia Earhart, women had found both a model and an advocate who became more eloquent with the passing years. At first she spoke out for women's role in aviation, stressing the serious contribution they could make and deploring the stunt flying that she thought reduced women to mere entertainers in a circus. In time, she became a champion of women's rights in general: She did not want to see women either pampered or penalized, she declared, adding that "sex has been used much too long as a subterfuge."

It was no accident that in the year after the *Friendship* flight, 20 racing planes set off from Santa Monica, California, for what Amelia Earhart called "the event that started concerted activity among women fliers." This was the Women's Air Derby, the first cross-country competition for women, staged as a major opening event of the National Air Races of 1929. The requirements for entry were a pilot's license and a minimum of 100 hours' solo flying time. Amelia Earhart estimated that only 30 candidates in the entire country were eligible for the race—a measure of women's difficulties in aviation. Almost two thirds of those eligible participated in the race.

AE herself was among the participants—as were most of the women who were to be the leading fliers of the 1930s. From overseas had come the German stunt flier Thea Rasche and Australia's Jessie Miller, who had recently flown as copilot on a five-month, 12,500-mile odyssey from London to Darwin. Among the American entrants were Ruth Elder of the abortive *American Girl* flight; former actress Blanche Noyes, whose mail-pilot husband, Dewey Noyes, had taught her to fly; skydiver and flying-school owner Phoebe Fairgrave Omlie; Marvel Crosson, a youthful veteran of commercial flying in Alaska; Hollywood stunt flier Florence "Pancho" Barnes; long-distance flier Ruth Nichols; California's Gladys O'Donnell, who was famed for her skill at closed-circuit racing; endurance flier Bobbi Trout; and Louise Thaden, who was the simultaneous holder of women's endurance, altitude and speed records.

The press called them Petticoat Pilots, Ladybirds and Flying Flappers, and humorist Will Rogers was on hand with wry comments at the start of the race that he and others had dubbed The Powder Puff Derby. A survey of male opinion at about the same time showed that women pilots were considered too emotional, vain, inconstant and frivolous—hazards to themselves and to others.

The Women's Air Derby disproved all that. An endurance marathon as much as a race, it was laid out in nine daily stages. The participants were to take off from Santa Monica's Clover Field on Sunday, August 18, and finish in front of the main grandstand at Cleveland Municipal Airport on Monday, August 26. Flying between sunup and sundown, they would cover 2,800 miles, or roughly 300 miles a day, with stops for rest and refueling.

It was not an easy course. For more than a week, the women "flew

low and wide open," as AE put it—finding their way by dead reckoning over deserts, mountains and plains. Their navigational aids consisted of magnetic compasses, sectional charts and Rand McNally road maps. In addition, each carried a gallon of drinking water and a three-day supply of malted-milk tablets and beef jerky to sustain her in the event of a crash in desolate terrain.

Waving triumphantly from her Travel Air J-5, a jubilant Louise Thaden, victor of the first cross-country Women's Air Derby, is wheeled into the winner's circle at Cleveland Municipal Airport. Only 23, she had already broken the women's international speed, endurance and altitude records.

Fatigue was the great hazard, Amelia Earhart recalled. Up each morning at 4 o'clock and in the air by 6, the contestants had to cope not only with the trials of the day's flying but with the interest of swarming

crowds. At fueling stops, the fliers found themselves signing autographs, answering questions and defending their planes from local flying enthusiasts who wanted to kick the tires or perhaps poke an umbrella through a wing to test its thickness. Yet few among the pilots lost their poise—either on the ground or in the air. On the fifth day, Blanche Noyes was flying at 3,000 feet over the western Texas desert when she looked behind her to see her baggage compartment being consumed by tongues of flame. She promptly landed—damaging a wheel in the thick mesquite brush—and put out the fire with handfuls of sand. Then she took off despite the damaged wheel—though she was never able to explain how she did it.

Other pilots were blown off course, ran out of fuel or were forced down by failing engines. Bobbi Trout cartwheeled during an emergency landing in a plowed field in California and amazingly emerged without a scratch, although her plane was badly damaged. Equally lucky was Pancho Barnes, who clipped a parked car on landing at Pecos, Texas, and walked away from the wreckage. Margaret Perry of Beverly Hills flew doggedly for two days with a high temperature before giving up in Fort Worth, Texas, where she was hospitalized with typhoid fever. Ruth Nichols got caught in a side wind that blew her plane into a tractor beside the runway at Columbus, Ohio, and put her out of the race.

But amazingly, despite the hundreds of hours that the women logged in their tricky planes, there was only one fatality in the race. In the bleak Gila River mountain country of southwestern Arizona, Marvel Crosson bailed out of her stricken airplane at too low an altitude and was killed instantly when she hit the ground, wrapped in the silk fabric of her partially opened parachute.

Public reaction was predictable. Fatalities in men's racing were seen as occupational hazards, but in women's racing they were signs of incompetence. "Women Have Conclusively Proven They Cannot Fly," declared a headline. And there were demands for cancellation of the derby—which the race committee rejected.

Of the 20 fliers who had started the race, 15 brought their planes into Cleveland. Louise Thaden was the winner, with Gladys O'Donnell second and Amelia Earhart third. Louise Thaden recalled watching the other planes come roaring across the finish line until "a glistening row of wings shimmered in the sun, lined up in front of the long center grandstand." It was a proud sight, she thought, but also a sad one: "We were all there, an undetermined, aimless group, now that the Derby had ended."

They were not aimless for long. A bond had been forged among women fliers, and soon it was formally recognized: On November 2, 1929, four of the Derby participants, along with 22 other women, met at Curtiss Field in Valley Stream, Long Island, to form an association of female fliers they called the Ninety-Nines.

The name was derived from the number of charter members: 99 licensed women pilots immediately joined up. Dedicated to the im-

The 99er

June, 1934

Vol. I, No. 8

IN THIS ISSUE

The Aerial Eye of 2,000,000 People
By H. R. McCORY

Selling Wacos

As Carolina Fathers Cheer

Flying Blind
By GORDON SILVA

COMING NEXT MONTH

The Inherent Characteristics of a Flier
By Dr. W. G. GAMBLE

Mexico's Airports
By BESSIE OWEN

Aviation in the Philippines
By FAY GILLIS

Tomorrow's Pilots

Mary Riddle of Seattle, only Indian woman flier, has an important message: "Come on, let's help fill the sky with thunder birds." (See page 3)

A MAGAZINE ABOUT WOMEN AND FLYING

A cover of The 99er, the first magazine produced by, for and about women fliers, features Mary Riddle, then the only licensed American Indian woman pilot. The 99er was published by the Ninety-Nines, an organization of licensed women pilots whose first president was Amelia Earhart.

provement of women's opportunities in aviation, the organization was largely responsible for making women an accepted part of the National Air Races and of competition flying during the 1930s. In time, it would become international and include virtually all the leading women fliers of the day.

At least four charter members of the Ninety-Nines were looking beyond domestic air competition to the challenge of the North Atlantic. Amelia Earhart, of course, was one. Another was New York-born Elinor Smith, whose courage and tenacity in seeking altitude records were legendary: Once she fainted at 25,000 feet when her oxygen tube broke; she fell more than four miles before regaining consciousness just 2,000 feet above the earth. Like Amelia Earhart, she was determined to fly the Atlantic solo, and she even made plans for such a flight. Eventually she was forced to abandon the project for lack of funds—as did Germany's Thea Rasche, who was confident of being the first woman across before she had a disagreement with the company backing her.

AE's most formidable rival for the first woman's transatlantic flight was the indefatigable Ruth Nichols, who impressed no less a figure than Commander Richard E. Byrd with her "indomitable pluck." No matter what happened, noted Byrd in wonder, Ruth Nichols "always got up and tried again."

She survived six major crack-ups in her years of competition, and she had innumerable close calls. No career in women's aviation—with the exception of Amelia Earhart's—was more inspiring to younger fliers as an example of perseverance against inhibiting odds. Although she had larger family resources than many other fliers—her father was a New York stockbroker—Ruth Nichols was also exposed to stronger pressures to abandon flying in favor of a conventional social career. Nevertheless, it was her father who inadvertently launched her into a flying career. As a reward for graduating from Miss Master's School in Dobbs Ferry, New York, he took her to an air show in Atlantic City, New Jersey, in 1919 and permitted her to go aloft for a 10-minute ride in an open-cockpit airplane. That was all she needed: "I haven't come down to earth since," she wrote in her autobiography many years later.

Instead of making her debut into New York society, as the family had planned, she delayed it for two years and went to Wellesley to begin preliminary studies in medicine. In Miami during a year's break from school, she went up for a ride in a seaplane and was given a chance to hold the controls. Exhilarated by feeling "the power of my own hands managing this fierce and wonderful machine," she started taking lessons from the pilot, a barnstormer named Harry Rogers. She won her license just after graduating from Wellesley but was dissuaded by her family from either going on to medical school or seeking a career in aviation. She took a job in a bank.

What catapulted Ruth Nichols into a flying career was an invitation from Harry Rogers to go along as copilot on a flight to Miami. She accepted. They made it in a record 12 hours, and Ruth Nichols found herself—as Amelia Earhart had after the *Friendship* flight—a sudden celebrity simply because she was a woman. Typical of the newspaper coverage of the event was a headline proclaiming "Flying Deb Pioneers New York-Miami Hop."

On the strength of the publicity, she got a promotional job with the Fairchild Airplane and Engine Company. From that time on—she was then 27 years old—she was able to make her living in aviation. With the foresight and single-mindedness that were characteristic of her, she started planning flights that would hone her skills for an assault on the North Atlantic.

In late 1930, in a borrowed Lockheed Vega, she broke the women's east-west transcontinental speed record and then turned around and broke the record for flying west to east. Then she stripped down the Vega; encased herself in long underwear, four sweaters, flying suit,

Amelia Earhart appears in a 1928 ad that capitalized on her sudden fame as the first woman to cross the Atlantic by air.

A 1939 advertisement featuring Betty Huyler Gillies, a well-known professional flier, points out that she was a mother as well as a pilot. The implied message was: If housewives could not aspire to Betty Gillies' aviation achievements, at least they could emulate her by lighting a cigarette.

Trees sway in the wake of Florence "Pancho" Barnes's speeding racer in this painting of her record-breaking speed trial in 1930. Union Oil often featured women pilots, among them Louise Thaden, in its advertisements.

Fastest mile ever traveled by a Woman…!

Florence Lowe Barnes, flying her Travel Air Mystery S, fueled with Union-Ethyl Aviation Gasoline, sets new world's speed record at 196.19 M. P. H.

ROARING over a measured mile course, only 164 feet above the ground, Florence Lowe Barnes made aviation history at Los Angeles Municipal Airport on August 4 when she set a new world's speed record for women.

It was a combination of sheer nerve, skillful piloting, faultless motor and ship, and perfect aviation fuel.

The speed averaged 196.19 miles per hour—*18.35 seconds for the mile course.*

The event was supervised and timed by Joe Nikrant, N. A. A. official.

Why She Chose Union

Mrs. Barnes knew that for maximum power, uniformity, and dependability Union-Ethyl Aviation Gasoline has no equal. She knew that it powered the Southern Cross on its epoch-making flight across the Pacific, and that it has a long record of unsurpassed performance.

We are glad that it met her expectations.

Use It Yourself!

Give yourself the extra security and other advantages that Union Aviation Gasoline offers. It is available to you at all leading western airports, or direct from the Aviation Department, Union Oil Company, Los Angeles, California.

UNION ETHYL Aviation Gasoline

UNION OIL COMPANY

Cashing in on sudden fame

When women pilots of the aviation-mad 1920s and 1930s set records or won races, they became instant celebrities and were besieged by advertisers with requests to endorse their products. Many fliers touted particular brands of gasoline or spark plugs—the tools of the trade. But some famous pilots endorsed goods having nothing to do with aviation.

Occasionally an endorsement backfired on the aviator. Amelia Earhart appeared in an advertisement for Lucky Strike cigarettes even though she was a nonsmoker. She pacified her conscience by not saying she actually smoked the cigarettes and by giving her $1,500 fee to Commander Richard E. Byrd for an Antarctic expedition. Yet many Americans were shocked by the association of their heroine with tobacco, and *McCall's* magazine showed its displeasure by firing her from her job as aviation editor.

boots and mittens; and on March 6, 1931, took off from Jersey City for an assault on the women's altitude record.

Her oxygen-supply system was notable for its simplicity: "We didn't have enough money to buy the right kind of set, so I just took a regular oxygen tank and attached a rubber hose and sucked the oxygen into my mouth directly from the tank." The trouble was that the exposed oxygen tank cooled off as fast as the outside temperature dropped: When the plane was at 20,000 feet, she was sucking up oxygen that had cooled to 60° below zero. The natural moisture in her mouth froze, numbing her tongue and causing her, as she later acknowledged, "a great deal of pain as well as a great deal of concern." But not concern enough to make her stop: She bored on up into a sky turning darker blue until she was down to her last five-gallon reserve tank of fuel. Then the engine stopped abruptly, and she dropped 5,000 feet like a stone, almost rupturing her eardrums. When she got back to the airport with an almost empty tank and was congratulated on setting a new women's altitude record of 28,743 feet, she found that her frozen tongue refused to form any words.

Conspicuous in her purple leather flying suit and purple helmet, Ruth Nichols next set a new women's three-kilometer speed record of 210.6 miles per hour, beating Amelia Earhart's old record by nearly 30 miles per hour. By 1931 she had flown higher and faster than any other woman in the world. She was ready for her Atlantic flight.

The Vega was rigged with extra fuel tanks and the latest in navigational equipment. The plane was named *Akita,* an Indian word meaning "to discover." On June 22, 1931, Ruth Nichols flew from New York to St. John, New Brunswick, a refueling stop on the way to Harbour Grace, Newfoundland, from which she was to depart for Paris. After four hours she was over St. John's airfield, which was set in a tiny bowl amid cliffs and hills.

Looking down at the truncated runway, she realized that she would have to drop steeply over the wooded hills and brake the heavy, fast-landing Lockheed to a stop almost as soon as she touched down. She misjudged, tried to soar into the air again as she ran out of runway and slammed into the flank of a hill.

From a hospital in St. John, she sent a reassuring telegram to her mother: ALL I DID WAS WRENCH MY BACK AND WRECK THE SHIP. EVERYTHING UNDER CONTROL. AWFULLY SORRY ABOUT CRACK-UP. WILL DO IT NEXT TIME. LOVE, RUTH.

In fact, she had broken five vertebrae and was told she could not fly for a year. She paid no attention. Immobilized in a plaster cast at home, she supervised repair work on the *Akita* and made plans for a second try. At the end of a month, she drove her car for 13 hours to prove she could endure a Paris flight.

By September she was ready to go. Still encased in plaster, she scanned the weather bulletins day after day, waiting for clear skies. For 25 consecutive days it rained in Paris. When September rounded

A serene and determined Ruth Nichols wears her custom-made purple flight suit for this magazine portrait published soon after her 1931 attempt to fly the Atlantic Ocean. Combining argument and entreaty, she had persuaded Powel Crosley, a Cincinnati industrialist, to allow her to use his Lockheed Vega for the flight.

Ruth Nichols' friend and mentor, Clarence Chamberlin (on the wing), checks the damage to her borrowed Vega after she crashed in New Brunswick, ending her transatlantic attempt. Hospitalized by the accident, the determined flier told her doctor, "I've got to get out of here right away. I've got to fly the Atlantic!"

into a stormy October, she reluctantly postponed the transatlantic venture and decided instead to attempt to beat French flier Maryse Bastié's world distance record of 1,849 miles with a nonstop flight from California to New York.

She flew in easy stages to California, had her plane refueled and checked at the Oakland airport, and took off for New York. Although she was no longer in a cast, she was wearing what she later described as a steel corset that reached from under her armpits to down below her hips. It was, she recalled, "just like being in a straitjacket." When she needed to take the weight off her thighs and lower spine, she reached up to a pair of straps suspended from the top of the cockpit, slipped her arms through the loops and hoisted herself upward for a few moments of blessed relief.

East of Chicago she ran into heavy crosswinds and a deepening haze that foiled every effort to make a visual position check. Unable to get her bearings, she decided to put down at the next airport—which turned out to be Louisville, Kentucky. She had not reached her goal, but she had flown 1,977 miles, beating Maryse Bastié's record by more than 100 miles.

On takeoff the following day, the plane's exhaust set fire to fuel from a leaking valve, and the fuselage was enveloped in flames. Although impeded by her steel corset and a back parachute, Ruth Nichols somehow managed to brake the aircraft to a stop, clambered onto a wing and leaped to safety just seconds before the gasoline tank exploded. The

Akita was a total loss—and so were her dreams of flying it to Paris.

The only serious candidate left was Amelia Earhart. In the months following the Air Derby, AE had set several speed records in her Lockheed Vega and had become a roving ambassador for aviation, extolling its safety and practicality to audiences all over the United States. For *Cosmopolitan* magazine, which hired her as aviation editor, she turned out monthly articles with such titles as "Is It Safe For You To Fly?" and "What Miss Earhart Thinks When She's Flying." The combination of promotional work, product endorsements *(pages 60-61)* and her writings—which included *20 HRS. 40 MIN.*, an account of the flight of the *Friendship*—gave her genuine financial independence for the first time in her life.

She savored her self-sufficiency and freedom. In 1930, when she was 33 years old, she wrote to a friend: "I am still unsold on marriage. I think I may not ever be able to see it except as a cage." Not a year later, she astonished her friends, her family and perhaps even herself by announcing that she was marrying George Palmer Putnam.

Ten years older than she was, Putnam had been a close friend ever since the flight of the *Friendship.* He advised AE about her writings, her lectures, the kinds of jobs she should undertake. He asked her to marry him for the first time in 1929, just after he and his first wife had separated—but he had to propose six more times before she agreed. When she finally accepted, it was with a casual nod and a quick pat on the arm as she climbed into her plane to take off on yet another business flight.

From the start, she made it clear that theirs was not to be a conventional marriage. Just before the ceremony she handed Putnam a remarkable declaration of independence and conditional love: "Dear G.P.," the document began, "you must know again my reluctance to marry, my feeling that I shatter thereby chances in work which means so much to me. I feel the move just now as foolish as anything I could do." Accordingly, she had fashioned an assortment of reservations: "In our life together, I shall not hold you to any medieval code of faithfulness to me, nor shall I consider myself bound to you similarly. I may have to keep some place where I can go to be myself now and then, for I cannot guarantee to endure at all times the confinements of even an attractive cage. I must exact a cruel promise, and that is you will let me go in a year if we find no happiness together. I will try to do my best in every way."

The odd partnership seemed somehow to work—perhaps because its terms were so clearly defined in advance. Putnam devoted himself to boosting his wife's flying career. He signed her up for so many promotional tours, in fact, that they did not have time for a honeymoon. With Putnam handling her affairs, her picture appeared on billboards advertising cigarettes and her name appeared on the labels of a line of women's clothes. She endorsed automobiles and pajamas and lent her name to a brand of lightweight airplane luggage that was an immediate success. She accepted the commercialization of her career as a way of raising money for serious flight. Occasionally, though, she grew weary of what she called "the zoo part" of her work. While making a cross-country promotional tour in an Autogiro with "Beech Nut" lettered on it, she wrote sardonically to a friend: "Here I am jumping through hoops just like the little white horses of the circus."

Over breakfast one morning in early 1932, she asked Putnam if he would mind if she flew the Atlantic. It was largely a rhetorical question, for she had already made up her mind. He enthusiastically agreed. By this time his wife had flown more than a thousand hours in all kinds of weather and had become adept at instrument flying. Her good friend Bernt Balchen, whom she regarded as one of the finest fliers alive, readily agreed to act as her technical adviser for the flight and take charge of reconditioning her plane.

The fast and tricky Lockheed Vega she was flying was known among pilots as a very exacting plane. Loaded with extra fuel tanks in wings and cabin, it became more exacting still. Under Balchen's supervision, it got a new engine and additional navigational equipment and had its fuselage strengthened. With a new cruising range of 3,200 miles, it seemed entirely adequate for the flight.

On May 19, 1932, clear skies were reported over the Atlantic, and AE took off with Balchen from New Jersey's Teterboro Airport for Harbour Grace, Newfoundland. The next day Balchen made final checks on the plane at the Harbour Grace airport while AE napped until it was time to go. "She arrives at the field in jodhpurs and leather flying jacket," Balchen recalled in his journal, "her close-cropped blond hair tousled, quiet and unobtrusive as a young Lindbergh. She listens calmly, only biting her lip a little, as I go over with her the course to hold, and tell her what weather she can expect on the way across the ocean. She looks at me with a small lonely smile and says, 'Do you think I can make it?' and I grin back: 'You bet.' She crawls calmly into the cockpit of the big empty airplane, starts the engine, runs it up, checks the mags, and nods her head. We pull the chocks, and she is off."

For the first several hours it was easy. Flying at 12,000 feet, AE marveled at the lingering traces of sunset and at the moon rising over a bank of clouds. Then suddenly she saw the hands of her altimeter spin around wildly; the instrument was broken and would be out of commission for the rest of the flight.

Buffeted by a sudden electrical storm, she flew through it and then tried to climb above more clouds that enveloped her on the other side of it. Soon she noticed that the controls were sluggish and her speed had dropped. Ice, she realized, was forming on the wings. Even as she was preparing to descend, the Lockheed suddenly plunged downward in a spin. At one point during the headlong spiral the barograph recorded an almost vertical drop of 3,000 feet. "How long we spun I do not know," Amelia Earhart wrote afterward. "I do know that I tried my best to do

New York City welcomes Amelia Earhart with a ticker-tape parade after her 1932 solo flight across the Atlantic Ocean.

exactly what one should do with a spinning plane, and regained control as the warmth of the lower altitude melted the ice. As we righted and held level again, through the blackness below I could see the whitecaps too close for comfort."

She continued to fly in sight of the cresting waves until they were obscured by fog. Then she climbed again, guessing at an altitude between water and ice. About four hours out of Harbour Grace she noticed blue exhaust flames leaking out of a broken weld in the engine manifold. The flames were not a present threat, but if they burned through the heavy metal manifold ring the plane would be finished. "I was indeed sorry that I had looked at the break at all," she would later observe drily, "because the flames appeared so much worse at night than they did in the daytime."

By dawn, the manifold was vibrating badly. Thin daylight fell upon the sea, and ice particles glistened on the plane's wings. AE, sipping tomato juice through a straw, wondered if she might have been blown off course or if the Lockheed had used too much fuel. She switched on the reserve tanks and found that a line leading to the fuel gauge was leaking.

With instruments out of order, manifold loose and gasoline leaking into the cockpit, she obviously was not going to make it to Paris. She turned due east, setting what she hoped was a course for Ireland. She was almost convinced that she had missed it when she saw the humped shape of a coast rising over the horizon.

Passing over the coastline, she looked for a place to land. Without an altimeter, she was afraid to fly too far into the inland hills, so she turned north, followed some railroad tracks and came at last to "lovely pastures." She landed in a long, sloping meadow amid badly frightened cows. She turned off the switch and was struck by the sudden silence. An astonished farmer appeared and told her that she was near the city of Londonderry. She smiled at him amiably: "I've come from America," she said.

The flight had taken 15 hours 18 minutes, the fastest crossing on record. When she was awarded the gold medal of the National Geographic Society by President Herbert Hoover after her return to the United States, AE made light of her achievement. "Some features of the flight I fear have been exaggerated," she said. "It made a much better story to say I landed with but one gallon of gasoline left. As a matter of fact, I had more than a hundred. I did not kill a cow in landing—unless one died of fright."

But Amelia Earhart knew full well what she had accomplished. Her flight vindicated all the courageous women who had tried, in one manner or another, to fly the Atlantic Ocean before her. In surmounting that formidable barrier, she had demonstrated that nothing in the field of aviation was beyond women's capacity to achieve. In that sense, all women fliers had shared in her victory when she set her plane down among the cows in a Londonderry field.

N·Y·
ELINOR
SMITH
RICHFIELD
AVIATION
GASOLINE

An endurance test beyond endurance

The year 1929 saw a hot competition develop between two young women aviators from opposite ends of the country. In January, 23-year-old Bobbi Trout of California set a new women's solo endurance mark of 12 hours, but within a month New Yorker Elinor Smith, 17, broke it by an hour. The Californian promptly raised the record to 17 hours, and in April, Elinor Smith again outdid her with a flight of 26 hours.

It was a neck-and-neck contest with no end in sight until a California businessman offered to sponsor them both if they would make an endurance flight together with aerial refueling—a feat never before attempted by women. The rivals teamed up in November and set their sights on the formidable, 420-hour world endurance record set by two men, Dale Jackson and Forrest O'Brine, in July 1929.

At Metropolitan Airport outside Los Angeles the women practiced refueling their plane, a Sunbeam, from an ancient Curtiss Carrier Pigeon—appropriately dubbed the "wet nurse" and flown by two male pilots. On November 16 the women took off, determined to stay in the air for at least a month.

But their flight had barely begun when they were compelled to land; a heavy radio they had installed for communications with the Carrier Pigeon and the ground had badly unbalanced their plane. They took off again without the radio, but then the Sunbeam's wire rigging started to snap, forcing another landing. On November 25 they tried again, managing a flight of 18 hours before encountering further—and potentially disastrous—trouble: A refueling mishap soaked Bobbi Trout with gasoline. She landed safely, though sputtering that she had swallowed the fuel, and spent the night recovering in a local hospital.

On November 27, 1929, the two went up once more. After some 30 hours, just as they were beginning to think their luck had improved, their aged "wet nurse" began to belch clouds of smoke. It made a forced landing in a nearby field, and when the Sunbeam ran out of fuel, the women had no choice but to land too. At last they gave up, but they had set a new women's endurance record of 42 hours—and in the process became the first women aviators ever to refuel a plane in mid-air.

In a rehearsal of their flight, Bobbi Trout demonstrates the use of a makeshift sleeping compartment atop the plane's fuel tanks while Elinor Smith sits alertly at the controls.

While Elinor Smith (standing) looks on, Bobbi Trout gets a vigorous massage from a local chiropractor to put her in shape for their long flight. Both women received such treatments, although apparently more for publicity than for reasons of health.

Elinor Smith adjusts the Sunbeam's enormous radio, while her partner hams it up with an outsized early microphone. The bulky vacuum-tube radio set proved too heavy for their fuel-laden plane.

At right, the women examine an aerial-refueling funnel on the side of their plane. They practiced connecting it to a hose in a hangar at Metropolitan Airport (below) before testing the system aloft.

Simulating the receipt of a parcel from the companion plane, Bobbi Trout stretches to reach a leather bag containing food, oil for the engine and mail. The bag was attached to a rope carried by the "wet-nurse" aircraft; the same rope was also used to raise and lower the gasoline hose.

The Sunbeam eases into position for refueling below the Curtiss Carrier Pigeon flown by pilots Peter Reinhart and Paul Whittier. The Curtiss could deliver 185 gallons of gasoline within four minutes.

REFUELING

IL MATTINO
ILLUSTRATO
Rd 5

3

Farther, faster, higher!

A slight, boyish-looking British woman with the unremarkable name of Amy Johnson strode across the tarmac of Croydon Aerodrome on the cold, misty morning of May 5, 1930, heading for a small open-cockpit biplane with *Jason* lettered on its engine cowling. She was an unknown but enormously ambitious pilot, and in her map case she carried the route for one of the most ambitious flights in aviation history.

She greeted the small knot of well-wishers—a few friends and her father—who stood huddled in heavy coats around the airplane. There was no crowd of watchers, as there usually was in the 1930s when more widely publicized aviators were about to take off on daring air journeys; because Amy Johnson was not famous, the British press had drawn little attention to the 26-year-old woman or her projected flight.

The modest send-off bothered Amy Johnson not at all. She flashed a cheerful smile, gave her father a kiss on the cheek and climbed into the plane's cockpit. She taxied down the runway, stopped some distance from its downwind end, turned and began her takeoff run. The aircraft, heavily laden with 60 gallons of extra fuel, had not yet lifted from the ground when she saw the boundary fence at the windward end of the runway looming up. She throttled down in time to avoid disaster and returned to the far end of the field. This time she roared past the few onlookers, skillfully coaxed the *Jason* off the ground and climbed over the suburban houses beyond the airfield boundary.

Once aloft, she set her course for Vienna, embarking on an 800-mile flight that would take a grueling 10 hours in the *Jason,* a de Havilland Gipsy Moth with a puny 100-horsepower engine. But that was only the first leg of the improbable air voyage Amy Johnson had in mind: Her ultimate destination was Australia, almost halfway around the globe—some 10,000 miles over forbidding stretches of sea, desert, mountain and jungle.

Amy Johnson was determined not only to make the flight but also to beat the record of 15½ days set in 1928 by a veteran pilot named Bert Hinkler, the first person ever to complete the trip flying solo. But what made her undertaking truly extraordinary was her lack of flying experience. Before leaving Croydon that May morning she had logged only about 85 hours of solo time, and her longest previous cross-country flight had covered precisely 147 miles, from London to her hometown of Hull on England's east coast. Yet she never doubted her ability to

Polish aviator Sofia de Mikulska, shown on this 1930 cover of a Naples weekly, is surrounded by buffalo on Italy's Garigliano River plain after an emergency landing. Forced landings in unfamiliar locations were frequent occurrences for long-distance fliers in the 1930s.

make the trip. "The prospect did not frighten me," she said later, "because I was so appallingly ignorant that I never realized in the least what I had taken on." But such modesty could not hide her gritty determination, her single-minded drive.

Before Amy Johnson took off on her solitary flight Americans had been leading the field among women aviators for several years. During the 1930s, however, a number of women pilots from Britain, other British Commonwealth countries and Europe competed on even terms with such outstanding American fliers as Amelia Earhart and Ruth Nichols. With one breathtakingly ambitious flight after another, they challenged scores of long-distance, endurance and altitude records both in powered aircraft and in gliders. Their triumphs were greeted with such outpourings of public enthusiasm that for a few golden years women fliers became international celebrities. They provided, with their examples of courage, some of the most luminous moments of a decade that darkened ominously with the gathering clouds of war.

This 1938 portrait of the once jaunty Amy Johnson—the first woman to make a solo flight from England to Australia—illustrates the cool sophistication that she cultivated after her troubled marriage to playboy-aviator Jim Mollison.

Amy Johnson was born in 1903 in the ancient port city of Hull, where her father was a fish merchant. She developed a competitive streak as a schoolgirl, playing a vehement brand of field hockey and challenging the local boys at the male-dominated sport of cricket. She entered Sheffield University in 1922, when it was still unusual for young women from Britain's provincial cities to seek higher education.

After earning a bachelor's degree, she spent an uncharacteristically indecisive three years drifting from one office job to another, ending up as a secretary to a London lawyer. But such a humdrum existence was not for Amy Johnson. In the spring of 1928 the sound of airplanes flying from a small field called Stag Lane on London's periphery suddenly revived childhood dreams of glory—she had once told a neighbor that she wanted to be a queen—and for the first time she thought of flying as a route to fame and freedom.

Although she was earning only five pounds a week as a secretary, she quickly plunked down an initial fee of more than six pounds to join the London Aeroplane Club, which was located at Stag Lane, and spent her savings taking flying lessons. At first she showed little sign of any particular aptitude for a career in aviation. Smooth landings eluded her; she tended to thump her plane down onto Stag Lane's grass runways. But she stuck with it, soloed after some 16 hours of instruction and impressed the examiners with her skill as she easily passed the test for a private license.

What was most unusual about Amy Johnson was her determination to learn everything she could about aircraft maintenance as well as flying, knowledge that would later prove invaluable. She talked the club's ground engineer, a gruff perfectionist named C. S. "Jack" Humphreys, into taking her on as an apprentice, and she often showed up at Stag Lane at 6 o'clock in the morning to help him dismantle, clean and reassemble aircraft engines.

In December 1929, permeated with grime and enthusiasm, she became the first woman in England to earn an aircraft ground engineer's license. Within a few weeks she made the astonishing decision to use her newly won skills—and make a reputation for herself in the world of aviation—by attempting a long-distance flight no woman had ever undertaken before. Australia seemed to her to be a good target simply because it was the most distant destination she could think of that did not require crossing oceans wider than the range of the small planes she had learned to pilot (the longest overwater leg of the journey would be a 760-mile hop across the Java Sea).

Her decision made, she meticulously plotted every detail of her planned route, down to how much extra oil a light plane's engine would require to stay lubricated in hot climates. At the same time she sought financing for the project—and suffered some stiff rebuffs. She first tried London's newspapers, which frequently gave famous aviators expense money in exchange for exclusive rights to the stories of their adventures. But Fleet Street's moguls had never heard of Amy Johnson, and they scoffed at the neophyte pilot's chances of completing her proposed trip. An Australian trade minister, approached next, offered only avuncular advice: "Better go to Australia by steamer, my girl. You would be foolish to try to fly there."

Such setbacks only sharpened her determination. She wrote letters to every wealthy celebrity she could think of, asking for help. After weeks of stubborn effort, she managed to persuade Lord Wakefield, an oil magnate, to contribute toward the price of the plane she needed and to arrange for fuel supplies along her projected route. Even then her father had to chip in half of the £600 cost of her secondhand Gipsy Moth. With a nice mixture of poetic inspiration and daughterly gratitude she named it *Jason,* for the ancient Greek hero who had traveled so far seeking the Golden Fleece—and in honor of her father's fish business, which used the Greek Argonaut as a trademark.

The first leg of the flight, from Croydon to Vienna's Asperne Airdrome, was largely uneventful. The only unpleasant feature, Amy Johnson later recalled, was the nauseating odor of gasoline that filled the cockpit whenever she hand-pumped fuel from the *Jason's* reserve tanks into the main tank. The next leg, from Vienna to Istanbul, was similarly routine, but from there on she headed into less traveled and far more dangerous territory.

The third day of the trip took her across Asiatic Turkey's lofty Taurus Mountains, with peaks reaching above 12,000 feet. She would have to fly between the peaks because the *Jason,* laden with its extra fuel tanks, could climb no higher than 11,000 feet. Patches of dense cloud obscured the mountainsides, and deadly obstacles loomed with alarming suddenness. "Mountains rising sheer on either side of me only a few feet from my wings," she recorded in her notes detailing the flight. "Rounding a corner I ran straight into a bank of low clouds and for an awful moment could see nothing at all. In desperation I pushed down the nose of the machine to try to dive below them, and in half a minute—which seemed to me an eternity—I emerged from the cloud at a speed of 120, with one wing down and aiming straight for a wall of rock." Banking instinctively, she evaded that barricade and, to her great relief, soon spotted a railway line cutting through a pass far below. After descending through gaps in the clouds, she followed the tracks, snaking her way through the remaining passes. Finally leaving the Taurus Mountains and Turkey behind, she came down in Aleppo, Syria, to call it a day.

The next leg, from Aleppo to Baghdad, also offered Amy Johnson more excitement than she had bargained for. When almost in sight of Baghdad's airport she flew into a violent windstorm that tossed the *Jason* around like a kite and then suddenly dropped the little plane to within a few feet of the churning desert sands. "Sand and dust covered my goggles," she later wrote, adding, "I had never been so frightened in my life." Taking the only course open to her, she force-landed into the gale at full throttle. The *Jason's* undercarriage strained at the jolt but held, and the plane slithered to a stop.

Working feverishly in the blinding sandstorm, the young pilot chocked the wheels with luggage and tool boxes and then sat on the *Jason's* tail to keep the plane from being blown across the desert. Hearing the howling of desert dogs—which, she had been told, could tear a human victim to pieces—she unlimbered a revolver she had brought along and waited out the storm for three hours. No dogs appeared, the wind finally abated, and having gotten the *Jason* into the air again, she reached Baghdad in the late afternoon.

There she allowed herself the luxury of having the *Jason's* engine overhauled by professional mechanics stationed at that remote outpost to service the planes of Britain's early commercial airline, Imperial Airways. She usually did the greasy, three-hour job herself after each day's flying, which left her an average of only slightly more than three hours of sleep a night before her early-morning takeoffs.

Two more long days in the air, covering a total of 1,500 miles, brought Amy Johnson to Karachi, the gateway to the Indian subcontinent, only five days after leaving London. The British newspapers that had ignored her departure were by now breathlessly reporting the "flying secretary's" every landing, and she became an instant heroine for having beaten the previous London-Karachi record, set by Bert Hinkler during his 1928 flight, by an amazing two days.

Her hopes of surpassing Hinkler's overall London-Australia record were soon threatened, however. Trying to fly nonstop halfway across India, from Karachi to Allahabad, she was held up by persistent head winds and found the fuel tanks nearly empty while still more than an hour short of her goal. Spotting a military parade ground outside the village of Jhansi, she set the *Jason* down as softly as she could. But the *Jason,* like most light planes of the era, had no brakes and kept rolling until it slammed into a post at the far end of the field, smashing a wing. A local carpenter repaired the structural damage, the village tailor sewed

At England's Croydon Aerodrome, Amy Johnson receives last-minute instructions before embarking on her historic solo flight to Australia in 1930.

up the torn fabric and the military post's British garrison provided some fuel. Although the repairs were completed overnight, the unscheduled stop had cost her precious time.

Worse troubles followed as Amy Johnson flew from India across the Bay of Bengal and over the jungle-carpeted lands of Southeast Asia. Here airports were rare and emergency landing strips were almost nonexistent. On the 13th of May, nine days out from London, she was unable to find Rangoon's race track—the nearest thing to an airfield in that part of Burma—and had to make her landing on a soccer field. The field turned out to be too short; the *Jason* streaked past a soccer goal and went nose-down into a ditch. This time a lower wing, part of the undercarriage and the propeller were smashed.

Fortunately the soccer field belonged to a Burmese engineering school. Faculty and students pitched in and swiftly duplicated the *Jason's* broken wing ribs and landing-gear strut while a local druggist mixed up a reasonable facsimile of dope to retighten the patched wing fabric. An extra propeller that had been lashed to the *Jason's* fuselage throughout the trip—just in case—was installed, and Amy Johnson tenaciously resumed her flight to Bangkok and then southeastward down the Malay Peninsula to Singapore. But she had lost two days because of repairs, and her lead over Hinkler had vanished.

Further perils lurked in the Dutch East Indies—today's Indonesia—whose islands she intended to use as the last steppingstones to her Australian goal. Trying to reach the airport at Surabaya on the island of Java, she encountered a fierce rainstorm that pelted into the open cockpit. Impenetrable black clouds hemmed her in. "Unable to go on," she recalled, and "equally unable to turn back or to stand still, I circled round and round." At last a clear patch appeared and she dashed through it, finally sighting Java, but the circling had depleted her fuel reserve and another forced landing was necessary, this time on a sugar plantation. "I landed in a field," she wrote, "in which someone had planted long pointed stakes to mark out the design for a new house." The *Jason* headed unerringly for the stakes, which ripped through the already piebald fabric of the lower wings. This time they were patched with the only material at hand—pink adhesive tape.

The next hazard was the Timor Sea. It measured 500 miles across and at the *Jason's* cruising speed of only 80 to 90 miles an hour that meant seven exhausting hours of flying against head winds, much of it out of sight of land. Fighting her fatigue, Amy Johnson grimly held on until, as she recalled, "a dark cloud on the horizon slowly assumed shape." The cloud turned out to be Melville Island, off the northern coast of Australia. A few more minutes of flying and Amy Johnson at last touched down at Darwin. It was May 24; her trip had taken 19½ days.

Having seen a few newspapers during her journey, she was aware that the world's press had been following her progress since her stop at Karachi, but she was nevertheless dumfounded to find thousands of cheering Australians stampeding across Darwin's airport to give her a tumultuous welcome. As far as Amy Johnson was concerned, she had failed by four days to break Bert Hinkler's record and there was nothing to shout about.

But she was the first woman to fly alone from England to Australia and the world was agog over the onetime secretary who had completed such a long and perilous journey. Australia rocked to the beat of a welcoming song that, following her own preference in nicknames, was titled "Johnnie's in Town." This local hit was soon drowned out by the strains of "Amy, Wonderful Amy," which swept through the British Commonwealth after her triumphant return to England. Publicity-seeking companies deluged her with gifts, including an automobile, a speedboat and a new de Havilland Moth. King George V made her a Commander of the Order of the British Empire. And London's *Daily Mail* made up for Fleet Street's earlier cold shoulder by signing her to a package deal: £10,000—$50,000—for the dilapidated *Jason,* some articles she would write and a string of public appearances.

Amy Johnson made these appearances for only one week before ducking the limelight and hiding out with friends. Her warm and artless speeches had invariably been crowd pleasers, but the young aviator had quickly tired of public adulation. She wanted to be accepted as a serious flier and not romanticized as the humble little secretary who, as one headline writer phrased it, had "Quit Her Dull Job to Seek Adventure In the Air."

Apparently in a genuine effort to escape the publicity that kept hemming her in, Amy Johnson was soon back in the air again. But, inevitably, the press swarmed about whenever she took off or landed. The papers mourned when she was forced down in Poland during an attempted flight across Europe and Siberia to Peking in January 1931. There was rejoicing, however, in the summer of that year when she and her former engineering teacher, Jack Humphreys, teamed up to fly a new de Havilland Puss Moth from London to Tokyo in 10 days, setting a speed record for light planes on that 7,000-mile route.

Even more newsworthy was her marriage, in July 1932, to one of the leading British fliers of the era, James Mollison. Public and press went wild over the romantic union of the Queen of the Air and her intrepid hero, who had recently set a record of his own by flying from England to Cape Town, South Africa, in 4 days and 17 hours. The newspapers soon dubbed them the Flying Sweethearts.

Yet the Mollisons usually did not fly together. Within a month of the wedding, Jim Mollison flew his Puss Moth across the North Atlantic, the first solo Europe-to-America crossing ever made. By the end of the year his wife had broken Mollison's own record for the England-to-Cape Town flight and capped this achievement with a record-breaking return flight. Jim retorted with the first east-to-west solo crossing of the South Atlantic in February 1933, racing from England to Brazil by way of West Africa in 3 days and 10 hours.

Competition was not the ideal ingredient for a perfect marriage. In the interests of harmony and shared adventure, the Mollisons planned an assault on the world's long-distance flight record. They acquired a new twin-engined de Havilland Dragon, which they named *Seafarer,* and took off from Pendine in South Wales on July 22, 1933, headed for New York on the first leg of a triangular trip that they expected would take them to Baghdad and then back to London.

An exhausted Jim was at the controls after 29 hours of shared flying time when the *Seafarer* ran short of fuel, forcing him to attempt a night landing on a small airfield outside Bridgeport, Connecticut. The big plane shot past the end of the runway and smashed into a swamp. Jim hit the windshield, cutting his head. Amy was shaken and bruised. Within a couple of hours, thousands of Bridgeporters had converged upon the *Seafarer,* ripping it apart for souvenirs.

Planeless and with their schedule in tatters, the Flying Sweethearts gave up the project. Jim Mollison shortly returned to England, but his wife stayed in the United States for some months. Her fame and her personal warmth conquered America's flying community, and she became friends with Amelia Earhart, whose contributions to the cause

Hoping to become the first husband-and-wife team to fly the North Atlantic, Jim Mollison and Amy Johnson pass over the Irish coast and head for New York in their de Havilland Dragon in 1933. Since the heavy plane needed a very long runway, they took off from a smooth strip of beach.

of aviation struck the modest British pilot as far more significant than her own achievements.

The Mollisons made another attempt at togetherness in the air in 1934, flying a fast twin-engined de Havilland Comet in a newly established England-Australia trophy race. The first phase of the flight was a triumph for them as they roared from England to Karachi in a record time of 22 hours. But during the next leg of the flight severe technical difficulties with their aircraft forced the Mollisons down and they withdrew from the race.

The marriage started to founder. Jim, always a flamboyant man about town, began to drink more heavily—and philander more openly—than usual, while his wife outshone him in flying exploits, winding up on the front pages in 1936 by flying the London-Cape Town route again, setting new records for both legs of the journey and for the round trip. In 1938 the Mollisons were divorced.

By that time Amy Johnson had quietly retired to the country, where she rested her bruised spirit by gliding and by writing about flying. She would never again attempt a long-distance venture; in a shrinking world

A week after their transatlantic flight (opposite) ended in Connecticut with a crash that left them slightly injured, Mollison and Amy Johnson—both in bandages—visit President Franklin D. Roosevelt (right) and his wife, Eleanor (left), in New York, accompanied by Amelia Earhart.

there were few opportunities left for pioneering flights. But she would keep on flying until the end of her life. It was a heart-and-soul preoccupation for her. "There is nothing more wonderful and thrilling," she once wrote, "than going up into the spaciousness of the skies in a tiny plane where you feel alone, at peace with everyone, and exactly free to do what you want and go where you will, and you need not come down to earth until your petrol runs out."

If it was remarkable that a London secretary like Amy Johnson could appear as if from nowhere and fly to Australia, completing one of the most ambitious and hazardous flights of the spectacular 1930s, it was downright astounding when a young woman music student from New Zealand capped that feat and added others to it.

Jean Batten was born in September 1909 and spent her first years in Rotorua, a small town on New Zealand's North Island where her father practiced dental surgery. She discovered a passion for music and when she went to boarding school practiced playing the piano so diligently that her father dreamed that one day she might become a concert performer. But by the time she reached the age of 10 she had begun to be fascinated by airplanes and long-distance flying.

Her interest was initially spurred by the first flight ever made from England to Australia, a month-long journey completed in 1919 by two Australian aviators who were brothers, Ross and Keith Smith. Then in 1928 two great flying achievements rekindled her interest in aviation. The first was Bert Hinkler's solo flight from England to Australia, the second a Pacific-spanning air journey from San Francisco to Brisbane, Australia, by the great Australian pilots Charles Kingsford-Smith and Charles Ulm and two American crew members. "I was deeply interested in these two flights," she wrote in her autobiography, "and when later Charles Kingsford-Smith flew over the Tasman Sea to New Zealand my enthusiasm for aviation increased and I decided to become a pilot." Her decision was reinforced when, during a visit to Australia, she was taken for her first airplane ride by the famous Kingsford-Smith. "Cruising about high above the Blue Mountains," she wrote, "I had felt completely at home in the air and decided that here indeed was my element."

This came as an unpleasant surprise to her father, who declared that flying was far too dangerous and flying lessons prohibitively expensive. Torn between two loves, Jean Batten committed herself to the new one; she sold her piano to finance her start in aviation.

Jean Batten's mother, after the initial shock, became her daughter's most enthusiastic supporter. They should travel to England, a center of flying activity, the two women agreed, and in 1929 they journeyed together to London. There they quickly discovered the London Aeroplane Club and Stag Lane Aerodrome, Amy Johnson's home field. Jean Batten began to haunt Stag Lane and shortly earned her private, or A, license in one of the Aeroplane Club's handsome silver-and-yellow Gipsy Moths. With a youthful optimism resembling Amy Johnson's, she immediately tried to obtain backing for a flight to Australia. Somewhat to her surprise, neither her relatives nor friendly fellow New Zealanders were interested in financing her.

Undeterred, she went back to Stag Lane, rolled up her sleeves to learn aircraft maintenance, earnestly studied meteorology and navigation, and piled up enough flying hours for her commercial, or B, license, meanwhile exhausting her savings and making frequent trips to a pawnbroker to pay for the extra instruction. Once armed with the impressive credential of a B license, she soon found a fellow pilot who had enough confidence in her skill to help finance a solo flight to Australia—in return for 50 per cent of any proceeds that might result. Climbing aboard a secondhand Moth, she set out from Lympne airfield near the Channel coast in April 1933.

This trip, and Jean Batten's next attempt as well, were doomed by bad luck. In Baluchistan, India (now Pakistan), on her first flight she was forced down by a sandstorm, made a blind landing and damaged her propeller. After obtaining a substitute propeller she continued on her way, but within 70 miles her Moth's engine broke down. "A connecting rod broke," she explained later, "and went bang through the side of the crankcase." Trying to make what aviators call a dead-stick landing—that is, with no power—she set the plane down on a road outside Karachi. The Moth veered and smashed into a stone fence.

On her next try, in April 1934, she got no farther than Rome. Following a more southerly route than Amy Johnson had, Jean Batten was flying from Marseilles to Rome when head winds slowed her progress. She finally ran out of fuel, she recalled, "at midnight in teeming rain and pitch darkness over the Italian capital." She managed to maneuver the silent machine into a small field on Rome's outskirts, emerging from the cockpit with only minor injuries. But damage to the Moth ended this second attempt to reach Australia.

Returning to London to make yet another start, she discovered that the press was beginning to poke fun at her tribulations. "Try Again, Jean," advised more than one snide headline writer.

And so she did, less than a month later. On the surface, her chances seemed as slim as ever. Her patched-up Moth was now nearly five years old. And because of her mishap in Rome she was taking off at a time, mid-May, when monsoons could threaten Southeast Asia.

Miraculously, much of this third flight was a dream come true, a sightseeing tour over beautiful sunlit landscapes. One leg of the trip, however—from Rangoon to Victoria Point on the southern tip of Burma—proved a nightmare.

The weather forecast in Rangoon had warned that the monsoon season promised to arrive sooner than usual. Jean Batten nevertheless took off into an overcast sky, wearing her light tropical flying suit for comfort in the sullen May heat and hoping that the prediction would turn out to be incorrect.

Her concern for her plane overcoming a passion for elegance, Jean Batten puts a white silk frock at risk to inspect repairs being made on her aircraft's greasy engine at Calcutta in 1934. The plane had developed a severe oil leak during her flight from England to Australia.

But she ran into severe squalls as she traveled southward and five hours out of Rangoon found herself enveloped in treacherous-looking rain clouds. There was no way around the huge tropical storm and her fuel supply was running too low for her to turn back. She had no alternative but to push on through the turbulent weather. "The rain thundered down on to the wings of my airplane like millions of tiny pellets," she said, "and visibility was so bad that the wing-tips were not visible and the coastline was completely blotted out. It was like flying from day into night."

She pushed the aged Gipsy Moth blindly through the downpour, listening intently to the sputtering of the damp engine. "Very soon the open cockpit was almost flooded, and my tropical flying-suit wet through." Then, through a freakish break in the storm she glimpsed the dark blur of the Burmese jungle beneath her. Diving beneath the curtain of cloud, she flew back and forth for an anxious 35 minutes before locating a clearing in the forest—which providentially proved to be the landing field at Victoria Point.

The worst—indeed, the only bad part—was over. On May 23, 1934, 14 days, 22 hours and 30 minutes after taking off from England, Jean Batten landed in Darwin. She had beaten Amy Johnson's time by more than four days. All Australia again went wild with enthusiasm. Her homeland of New Zealand, which she visited by ship because her Moth lacked the range to cross the 1,200-mile-wide Tasman Sea, gave her an almost hysterical welcome. The streets of Auckland were packed with cheering crowds, and the New Zealand government ceremoniously presented the country's favorite daughter with a check worth $3,000. For six weeks she toured her native land, cheerfully enduring all manner of entertainments and making some 150 speeches. Unlike Amy Johnson, Jean Batten thrived on being in the public eye.

The career of this ebullient and much-adored young aviator almost came to an abrupt end on her return flight to England. About 250 miles out of Darwin, while flying over the lonely Timor Sea, the engine of her overhauled Gipsy Moth inexplicably coughed, then faltered, then choked into dead silence.

A terrible feeling of helplessness swept over Jean Batten as the plane started a slow, inexorable slide down through a cloud layer toward the dark expanse of water. Praying that the problem was merely a temporary fuel-line blockage, she gave the engine full throttle. There was no response. As if mesmerized, she watched the altimeter needle spin from 6,000 to 5,000 to 4,000 to 3,000 feet. "Surely this can't be the end!" she later remembered thinking. "No, it's impossible—there must be—some way out."

The only way out, it appeared, was to ditch the plane in the sea. "Undoing my shoes and flying suit, I reached for the small hatchet which I carried in case of emergency and placed it in the leather pocket at my side. There seemed a desperate chance that if I were able to land the machine on an even keel I might be able to cut one wing away and float on it."

Suddenly, just as the Moth was about to hit the water, its engine burst into life with a noise, she recalled, "that was nearly deafening in the stillness." Scarcely daring to take a breath for fear of breaking the spell, Jean Batten gently maneuvered the aircraft back up to 6,000 feet and held it there until she spotted Kupang on the island of Timor and descended to a safe landing.

The elderly Moth's balky engine sputtered, stopped and then restarted itself several more times before Jean Batten finally touched down at Croydon Aerodrome. Despite engine trouble she had made the trip from Darwin in 17 days and 15 hours and had completed the first England-Australia round trip by a woman. Britain treated her as a celebrity and she enjoyed being one, but already she was planning new record-breaking flights.

The first was to be a crossing of the South Atlantic from West Africa to Brazil. Only Jim Mollison had made that ocean trip before, and Jean Batten wanted to better his time of 17 hours and 40 minutes.

On her 26th birthday, in September 1935, she took delivery of a brand-new Percival Gull, a small, swift, closed-cockpit monoplane that had been fitted with auxiliary fuel tanks. Two months later she was in the Senegalese town of Thies, near Dakar, making final preparations for her flight to Natal, Brazil, on the far side of the Atlantic.

In the course of her previous flights Jean Batten had begun to build a dual reputation: She was a pilot who made meticulous preparations, leaving nothing to chance; she was also a woman who wanted to look her best at the social functions that inevitably followed a notable flight. At Thies, under the fascinated gaze of the airfield's French colonial commandant and a group of French air-force mechanics, she showed both sides of her personality. First, she insisted on doing the work on the engine herself and on supervising the refueling operations, making sure that all the fuel was carefully strained through chamois to remove even the smallest impurities. She stayed on the job, despite muttered complaints from the mechanics concerning what they considered her unfeminine behavior, until she was positive that the little Gull was in perfect running order.

Next, she sorted out her flying kit, discarding such heavy items as spare engine parts, a flare pistol, a tool kit and water drums. She had decided, she said, to leave everything but absolute necessities at Thies. Then, as the bemused commandant looked on, she plucked two evening dresses from the pile of supplies, carefully refolded them and stowed them in the plane's locker. Many years afterward inquiring reporters would still ask why she had made such a ridiculously feminine

Seated in the shade of her Percival Gull, a perky Jean Batten awaits help after a gasoline leak forced her down at Araruama, Brazil, while she was making a side trip after her record-breaking South Atlantic solo flight. She was soon rescued by pilots from the Brazilian Army.

choice of "necessities." Jean Batten's answer was brisk and practical: The dresses weighed practically nothing. If her flight was successful, she would need them. If she fell into the Atlantic, heavy equipment would be of little use.

As it turned out, she needed her party wear and not her extra equipment, although several times during an uneasy trip through storm clouds and drenching rain she wished that her funds had been sufficient for a radio she could have used as a navigational aid. But even without it, her navigation was perfect and she made her landfall near Natal with no trouble. With an overwater flight that lasted only 13 hours and 15 minutes, she had beaten Jim Mollison's time across the South Atlantic by more than four hours. Her total elapsed time from England to Brazil—a strenuous 5,000-mile trip—had been 61 hours and 15 minutes, a world record for anyone in any type of airplane and almost a day less than Mollison's England-Brazil time. After the flight was over she summed up her feelings: "I experienced once again the greatest and most lasting of joys: the joy of achievement."

The joy would be hers many times in the years that followed. In 1936, determined to fulfill what she considered "the ultimate of my ambition" and to prove the practicality of an England-to-New Zealand air route, Jean Batten piloted her Percival Gull from Lympne to Auckland in 11

Jean Batten attends a 1936 Paris ceremony arranged by the prestigious Aero Club of France to applaud her historic South Atlantic flight. She was honored by France's most illustrious aviators, including (on her right) Louis Blériot, the first to fly an airplane across the English Channel.

days, including a two-and-one-half-day weather delay in Sydney, Australia. For the first time, England had been linked directly with New Zealand, a major triumph for any pilot but particularly for one flying solo in a single-engined plane weighing less than 3,000 pounds. Not until 1980, 44 years later, did anyone better Jean Batten's record for a solo flight in a small plane from England to New Zealand. When that pilot—Britain's Judith Chisholm—landed in Auckland, she was greeted by the woman whose record she had broken, Jean Batten, who had made her epochal flight 16 years before her challenger was born.

The desire to set or break distance-flying records spread throughout the world during the remarkable 1930s. Among the aviators of Europe to be inspired by the examples of the British and American long-distance fliers were two Frenchwomen who had begun their careers in aviation doing quite different sorts of things. Marie-Louise Hilsz—better known by her nickname Maryse, a shortened version of Marie-Louise—started as a professional parachute jumper, evoking gasps from crowds all across Europe, at 500 francs a jump. But in 1930 she acquired a plane—yet another Gipsy Moth—and set out to fly 15,000 miles from Paris to Saigon in French Indochina (now Ho Chi Minh City, Vietnam) and back. She was delayed by a succession of mechanical troubles and the journey took her three agonizing months, but she was the first woman ever to make the round trip. After a flight from France to Japan in 1933, she tried the Paris-Saigon route again, this time in a larger and more powerful aircraft, and succeeded in reaching Indochina in a record five days and nine hours.

Her compatriot Marie-Louise Bastié—she, too, was called Maryse—began her career by seeking, and setting, endurance records. In 1929 she broke the women's solo record for nonstop flight by circling over Paris' Le Bourget airfield for 26 hours and 48 minutes, and the next year she stuck it out for almost 38 hours, a world record.

But the endurance flier soon became fascinated by the challenge of long-distance air journeys. On June 28, 1931, she left Le Bourget in a small, 40-horsepower Caudron that was carrying so much extra fuel it was virtually a flying gas tank. She headed across northern France and Germany, bound for the steppes of Russia. Steering by compass across a largely featureless landscape, she passed over Moscow the next morning and finally came down near Nijni Novgorod (now called Gorki). She had flown 1,849 miles—farther in a nonstop straight run than any other woman, and farther nonstop in a light plane than anyone else at all.

Now that she was a famous flier, Maryse Bastié could make a living from aviation. She did public-relations work for the Potez aircraft-manufacturing firm for some two years, then taught at a French flying school. Long-distance flights still fascinated her, however, and in 1934 she flew from Paris to Tokyo and back, becoming the first woman to make the round trip solo. Her competitive spirit flamed again when she

En route to New Zealand in 1936, Jean Batten's plane (lower left) sits on a runway at Sydney's Mascot Aerodrome. Cheered by the enormous crowd, the flier, who had just arrived from England in a record-breaking eight days, reminded the Australians that "the flight is not finished yet."

learned in 1935 of Jean Batten's record-setting flight across the South Atlantic. Not having enough money to purchase a suitable aircraft, Maryse Bastié at length persuaded the Caudron company to let her borrow one of their newest, fastest monoplanes and took off from Paris for Dakar in September 1936. From Dakar she made a beeline for Natal. Her transatlantic time was 12 hours and 5 minutes, more than an hour faster than Jean Batten's crossing. During World War II Maryse Bastié served in the Free French Air Force. She was killed in 1952 in the crash of a plane being flown by another pilot.

Like Maryse Bastié, an Englishwoman named Beryl Markham became interested in long-distance flying only after she had made her mark in a very different sort of aviation. Born in 1903, she moved with her family to British East Africa (now Kenya) when she was four years old. There she became an ardent horsewoman and successful trainer until, in 1930, she discovered the wonders of airplanes. Within 18 months she had logged 1,000 hours in the air and had earned her commercial license. Thereafter she made her living as a bush pilot, flying night and day out of Nairobi to the hinterland outposts of East Africa. She carried mail, ferried passengers and rushed medical supplies

Beryl Markham's Vega Gull rests nose-down in a Cape Breton bog after the first solo east-to-west North Atlantic flight by a woman.

to isolated mining settlements and hunters' camps. She also delivered provisions to safaris and then served as an aerial game spotter for such noted American sportsmen as Alfred Vanderbilt and Winston Guest.

For variety, Beryl Markham flew the East Africa-to-London route four times. In 1936, while on her third visit to England, she decided to fly the North Atlantic from east to west. Because of prevailing head winds, the east-west crossing was far more hazardous and took longer than the west-to-east flight. Her goal was to set a new light-plane record for the trip. Beryl Markham, who by her own calculation had flown a quarter of a million miles while working at her trade, saw no reason why she could not manage a straightforward flight from England to New York in a new Percival Vega Gull despite head winds, rainy autumn weather and the lack of a radio to help keep her on course.

Taking off from Abingdon, England, on September 4, she flew west with the night. The wind rose and the rain fell, and for 19 hours she flew blind through darkness and storm. When at last she saw daylight around her and water below, she felt confident that she could land at Sydney Airport, on Cape Breton Island, Nova Scotia, to refuel for the final hop to New York.

But as she approached the coast of Nova Scotia, her plane's engine began to sputter and cut out intermittently. Assuming that the problem was the result of an air-lock in the last of her reserve gas tanks—a bubble of air that impeded the flow of fuel—she tried to clear it by turning the petcock handles of all the tanks until her hands were bleeding onto her maps and clothes. It did not help, and the sputtering engine cut out completely when the Vega Gull was still several miles from Sydney Airport. The plane began a dead-stick glide down toward an expanse of black earth studded with boulders.

She described, in her autobiography, what happened next. "The earth hurries to meet me, I bank, turn, and sideslip to dodge the boulders, my wheels touch, and I feel them submerge. The nose of the plane is engulfed in mud, and I go forward striking my head on the glass of the cabin front, hearing it shatter, feeling blood pour over my face.

"I stumble out of the plane and sink to my knees in muck and stand there foolishly staring, not at the lifeless land, but at my watch.

"Twenty-one hours and twenty-five minutes. Atlantic flight. Abingdon, England, to a nameless swamp—non-stop."

To end an attempted flight to New York nose-down in a Nova Scotia bog was not a professional aviator's idea of unqualified success. Yet piloting a small single-engined plane from east to west across the North Atlantic against the relentless wind was no mean feat, and Beryl Markham was the first woman to achieve it.

For German women, it was not easy to even attempt record flights during the years between the Wars. Under the terms of the Versailles Treaty, Germany was prohibited from having aircraft that might conceivably be employed for military purposes. Fliers were generally restricted to gliders or small sport planes, and they were required to limit the length of their flights.

Thea Rasche, the first woman to earn her pilot's license in postwar Germany, satisfied her yearning to fly by stunting and racing sport planes in competition with the finest fliers of her time, in the United States and France as well as in her own country. Elly Beinhorn, in a minuscule 80-horsepower Klemm monoplane, managed to fly in short stages from Germany to the Far East in 1931 and from Germany to Australia in 1932. From there she and her plane took ship for South America, where she flew over the Andes Mountains, dodging between the peaks that her small plane could not fly over. Although others had made the mountain crossing before her, including French flier Adrienne Bolland in 1921, it was still a hair-raising trip. She topped this adventure with a flying circumnavigation of the African continent.

A third remarkable German woman pilot, Marga von Etzdorf, born into an aristocratic military family in 1907, managed for a time to fly larger and heavier aircraft even during the period of restriction. She earned her license in 1927 and quickly got a job as a copilot with Luft Hansa. She therefore had the opportunity, unusual for German pilots of either sex, to handle large aircraft, and to regularly fly a commercial route that covered considerable distances, from Berlin to Basel, Switzerland, via Stuttgart.

She soon was possessed by an urge to fly even longer distances. Buying a small Junkers Junior with an 80-horsepower engine, she learned aerobatics and then embarked in 1930 on a leisurely flight from Berlin to Istanbul, making several stops in the Balkans. She next undertook a longer trip that involved overwater jumps, from Germany to the Canary Islands. Although her flight ended with an accident in Sicily—she suffered the flier's humiliation of having to get home by train—this journey prepared her for her most successful long-distance effort, an 11-day solo flight from Berlin to Tokyo in 1931.

By this time one of Germany's most celebrated women pilots, Marga von Etzdorf became an unofficial ambassador of German aviation, giving lectures about her Tokyo flight in a half-dozen countries. Then, eager to try another long-distance journey, she set out from Germany for Australia in 1933. All went well until she tried to land in Aleppo, Syria, at the end of the second leg of the flight. Swirling winds and shifting sand at the Aleppo airport caused her to misjudge her approach and she crashed into the barrier at the end of the field.

She was not seriously injured, but her plane was badly damaged and her dream of a triumphant, record-breaking flight was shattered. The humiliation was too severe. Feeling that she had disgraced the noble Etzdorfs and her fatherland, she retired to a room attached to the dining hall at the Aleppo airfield, ostensibly to rest, and shot herself. As one of Germany's most noted pilots, she was given a state funeral by order of Adolf Hitler.

Few German fliers had Marga von Etzdorf's opportunities to pilot

Three of Germany's most famous women pilots—from left, Liesel Bach, Thea Rasche and Elly Beinhorn—relax before the 1931 Deutschlandflug, a cross-country race open to men and women. Thea Rasche did not compete because she had no plane of her own at the time.

powered aircraft; most had to make do with gliders. Among the finest glider pilots—and one of the most expert female aviators of all time—was Hanna Reitsch, a small blond woman with bright blue eyes and a broad smile, whose 1930s flights in motorless sailplanes took her thousands of feet aloft and kept her skimming the air currents for as long as 11 hours at a time.

Hanna Reitsch was born on March 29, 1912, in Hirschberg, Silesia, now part of Poland but German soil at the time. In one of her several books she recalled that even as a little girl she had ached to soar like "the storks in their quiet and steady flight, the buzzards, circling ever higher in the summer air." Her longing to fly grew, she recalled, "with every cloud that sailed past me on the wind, until it turned into a deep, insistent homesickness, a yearning that went with me everywhere and could never be stilled."

Like some other women pilots, Hanna Reitsch had to overcome stubborn parental resistance before getting her wings. It was tacitly accepted in the Reitsch family that a girl's task in life was to marry and have children. Still, her father, an eye surgeon, could not help encouraging her when she expressed interest in a medical career. He was less enthusiastic, however, when she decided, at the age of 13 or 14, that she wanted to be a missionary doctor—a *flying* missionary doctor. Dr. Reitsch compromised: If she would say not another word about flying until she had received her high-school diploma, he would reward her with a training course at the nearby Grunau school for glider pilots.

Saying nothing more about her real ambition, the young Hanna studied hard and passed her school examinations. Her father, choosing to forget their pact, then gave her an antique gold watch. She quietly returned it to him, reminding him of his promise. When further parental ploys still failed to deflect his daughter, Dr. Reitsch gave in at last and permitted her to take gliding lessons.

In the early 1930s the Grunau Training School was a center of German gliding activity, and its head, Wolf Hirth, was much respected in the German flying community of the time. To impress him was a passport to a future in aviation.

Hanna Reitsch found that her first task was to win the acceptance of her fellow students. She was the only female member of the class and she was still a tiny teenager, measuring half an inch over five feet and weighing less than 90 pounds. To the jeering young men in the class, this featherweight girl was bound to fail.

In an effort to impress her classmates, she began too boldly, taking a glider into the air when she had been instructed merely to make practice slides down the takeoff slope. The glider wavered precariously and dropped to earth. This premature performance occasioned much laughter and unwanted advice from the boys, and a three-day grounding for disobedience.

While languishing on the ground, however, she watched every slide down the takeoff slope and every brief flight with unswerving attention, and she listened closely to every word spoken by the instructors, mentally rehearsing all the right moves. Her powers of concentration were such that, almost as soon as her punishment was over, she could outperform her classmates with ease. She passed her first test flights with such brilliance that the great Wolf Hirth himself took over the rest of her glider training.

Several days after she had passed her final test, Hirth invited his star pupil to take up the school's newest glider. It was a great honor; only Hirth and the other instructors had previously flown the craft. Now a very small woman of 19 was being told that she could stay up as long as she liked—and wind conditions would allow.

"For the first time," she later wrote, "I was now free to fly without restrictions and I took off with feelings of real pride to soar for as long as the winds would blow, drawing on the loveliest songs I could remember and singing them out loud into the sky, hardly noticing that it rained and snowed, and coming down only when after five hours the wind finally died down."

When she touched down, she was greeted by an excited throng of people and congratulated on setting a women's glider endurance record with her five-hour flight.

Largely to please her father, Hanna Reitsch enrolled in medical school in Berlin, but she found that she remained more interested in the anatomy of aircraft than of human beings. Then, in May 1933, when she was back home in Hirschberg for the holidays, she experienced a dazzling if terrifying flight that spelled the end of her medical studies. One warm, cloudless day Wolf Hirth stopped her on the street and unceremoniously told her that she was to take up a Grunau-Baby, the very latest type of training glider. She hurried to the flying field and, still wearing her light summer street clothes, was soon sitting at the controls of the extremely fragile-looking craft, studying the dials on its instrument panel.

The flight began uneventfully, but after a few minutes the glider began to rise abruptly. Gripped by strong thermal updrafts, it shot skyward—from 3,000 feet to 4,000, to 5,000 and still higher. "And then, a million drumsticks suddenly descended on the glider's wings and started up, in frenzied staccato, an earsplitting, hellish tattoo, till I was dissolved and submerged in fear. Through the windows of the cabin, which were already icing up, I could see the storm cloud spewing out rain and hail."

The glider kept climbing until, at 9,750 feet, the instruments froze and the needles on the dials stuck before the pilot's unbelieving eyes. And the glider no longer answered to the controls as it swooped and bucked at the mercy of the vicious storm. All Hanna Reitsch could do was hang on, her bare hands turning blue from the cold as she sat there in her summer dress.

All of a sudden, she became aware that it was warmer and brighter—and that the earth was now overhead rather than below. The storm had

Marga von Etzdorf, the first German woman to pass licensing examinations for commercial, glider, sports and stunt flying, stands beside one of the many types of aircraft she piloted. In addition to flying, she lectured on aviation in Europe and Asia.

spewed out the glider, but not before flipping it upside down. Seizing the control column, she righted her craft and wafted gently down into a level pasture several miles from where she had taken off. Once back home in Hirschberg, Hanna Reitsch learned that she and the Grunau-Baby, shot heavenward by the tempestuous winds, had set an unofficial gliding altitude record.

A few months after this unplanned bit of record shattering, she was invited by Walter Georgii, a 44-year-old professor at the German Institute for Glider Research, to join an expedition to study thermal conditions in South America. Hanna Reitsch was delighted and earned the 3,000 marks she was required to contribute to the expenses of the project by doing some stunt flying in a romantic film called *Rivals of the Air,* whose heroine was a glider pilot.

In South America in early 1934 she found herself, as usual, the only female pilot in a group of men; but in Brazil and Argentina this proved to be an advantage. Attending receptions and performing aerobatics for delighted crowds when she was not occupied with research flights, the little 21-year-old German woman with the big nerve and enchanting smile was an instant hit. She also demonstrated her matchless skill by making a long-distance soaring flight over the pampas of Argentina; for this feat the Argentinians awarded her their Silver Soar-

Flanked by an official honor guard, Marga von Etzdorf lies in state in Hamburg in July 1933, having shot herself after a crash in Syria. It was believed that her suicide was prompted by a French officer's humiliating accusations of incompetence.

ing Medal, which had previously been won only by male glider pilots.

Back home in Germany she succumbed to Professor Georgii's insistence that she work with him at the Institute for Glider Research at Darmstadt. She stayed for 11 years, taking brief periods of time out to accumulate a rich variety of experiences. A few weeks after joining the institute, she set a new women's world record for long-distance soaring, covering more than 100 miles. A few months later, when still only 22 years old, she was invited to attend the Civil Airways Training School at Stettin, where she practiced cross-country flying and advanced aerobatics in twin-engined aircraft—and where, once again, she was the only woman in a class of men who quickly grew to respect her.

Yet her most important contributions to the cause of aviation at that time were made through the institute. Her work there consisted of test-flying new types of both motorless and powered aircraft, and of retesting existing types with a view to improving their performance. In 1936, after a rash of fatal crashes among glider pilots, Hanna Reitsch was given the assignment of testing a newly developed braking device intended to increase a sailplane's stability and set a limit to its maximum speed even in a vertical dive. These dive brakes, which resembled landing flaps, could be extended or retracted as needed by the pilot through controls in the cockpit.

On the first test, the turbulence set up by the dive brakes as she made a shallow dive from 13,000 feet shook the entire craft so violently that the control column was ripped out of her hands. She knew that if such a severe wing flutter were to occur during a vertical dive, the glider could break up in mid-air. Obviously, the device needed improvement.

Day after day, week after week, Hanna Reitsch and her colleagues devised new adaptations and made tests. Gliding at altitudes between 14,000 and 19,000 feet, she experimented with the new dive brakes until she felt that the final phase had arrived—it was time for a test in a vertical dive. With only a fleeting thought of what could happen should the brakes not hold down the glider's speed during the headlong plunge, she aimed straight for the ground. She plummeted thousands of feet, seeing the earth come closer and closer; and as she dived, the machine remained steady, its speed restricted to some 125 miles per hour. At 600 feet, she pulled out of the plunge, retracted the dive brakes and floated to a landing.

Fellow pilots, institute directors and mechanics came running up, overjoyed at the success of their efforts. In fact, the development of the new dive brakes was an important milestone in the history of aeronautics. A year later she had the satisfaction of flying a powered military aircraft that was fitted with the brakes she had tested.

That same year, 1937, marked the beginning of a new chapter in Hanna Reitsch's life. In spite of restrictions imposed by the Versailles Treaty, Germany had secretly built an air force. Recently this force, the Luftwaffe, had been unveiled to the world, and Hanna Reitsch was asked to serve as a test pilot at the Luftwaffe's testing station near

With a vigorous handshake and good-natured smile, glider pilot Hanna Reitsch greets Luftwaffe Colonel Alfred Mahnke in 1936 at the Rhön Soaring Contests. Her enthusiasm and expertise earned her the honorary title of Flugkapitän—flight captain—from Hitler in 1937.

Rechlin. She eagerly accepted this opportunity to fly every type of military airplane in the Luftwaffe's growing inventory. During the same period she became the first woman to pilot a helicopter, putting an experimental craft through a series of test flights that she capped with a public demonstration inside the Deutschlandhalle, Berlin's giant enclosed stadium. Following a vaudeville show that featured "Dancing Girls, Fakirs, Clowns and Blackamoors," her unprecedented hovering flight seemed so simple and unspectacular that the public scarcely comprehended the significance of this early vertical-lift machine or the virtuosity of its pilot.

But technically minded people, especially those who saw the military implications of the helicopter, were enormously impressed, and some, in other countries, were doubtless dismayed by this proof of Germany's enormous strides in the field of aviation.

There would soon come a period in Hanna Reitsch's life when she would be cut off from the camaraderie of her peers in other lands. Before that happened, however, she traveled to the United States in 1938 with two other German pilots to take part in the National Air Races in Cleveland. Like many another foreign visitor, she was stunned by the

towering reality of New York's skyscrapers, which she judged to be as high as "I had flown for my pilot's certificate." And she was taken aback at the Cleveland air races to see the American flag hoisted each morning by "beauty queens in bathing costume—an almost sacrilegious procedure to the German mind."

But the warm, spontaneous young German was captivated by the friendliness of Americans, and as a woman pilot who had fought her battles with jeering men she was charmed to find that the American

Hanna Reitsch takes off in her specially built Sperber Junior glider at the Rhön meet of 1936. The only woman among 61 entrants, she placed fifth in the plane "tailored to fit me so exactly, that once in the pilot's seat I could hardly move."

husband, unlike his German counterpart, often "carries the wife's shopping bag" and "helps with washing the dishes." She liked American men, she wrote, "for their spontaneous chivalry and for their lack of aggressive self-assertion."

What Hanna Reitsch did not note was that such domestic chivalry often failed to carry over into the world of aviation, and that even the best American women fliers were still not receiving the recognition, opportunities and rewards that their skills deserved.

The Duchess of Bedford prepares to snip a ribbon attached to a Gipsy Moth, which is about to take off to mark the opening of the air meet. Although the program cover (left) shows wingwalking, none was performed.

Flying into male flak

When Britain's female aviators staged their First All-Women's Flying Meeting, at Sywell Aerodrome in 1931, they were amply reminded that women pilots faced a special problem as great as any posed by weather, mechanical failures or competitors: the condescension of men. "I have never seen so high a proportion of good looking and well-dressed young women at any aeronautical function," wrote one male journalist, who noted that the meet was, "very appositely of a more emotional nature than usual."

And when he could bring himself to report on the event itself rather than the comeliness of the participants, he asserted that the ladies in the main race so unnerved watching male pilots that the latter took refuge under automobiles and later stampeded the airdrome's bar to restore their courage. Another observer elaborately praised one of the women for landing a sputtering plane without losing her head and mowing down the spectators. It was, he wrote with spurious heartiness, "a truly stout effort."

Despite the presence of what one magazine acknowledged to be "critical male humanity on all sides," the meet was a success. The participation of prominent social leaders such as the Duchess of Bedford *(left and below)* boosted public interest in women's aviation. More important, the races, stunts and mock bombing raids proved that women could handle aircraft with daring and skill.

Fliers and organizers of the meet gather around the Duchess of Bedford, who holds a bouquet dropped from the Gipsy Moth that opened the flying show. They include (from left) a pilot, the Honorable Mrs. Victor Bruce; Molly Olney, who helped to organize the event; the Duchess; and the Countess Drogheda.

Watched by the crowd, a businesslike Dorothy Spicer, in helmet and overalls, spins the prop of the Spartan she shared with pilot Pauline Gower.

At the finish of the Ladies' Race, Susan Slade's Moth wings in first (left), followed by the Honorable Mrs. Victor Bruce in her Bluebird (right).

4

The woman who wanted "just one more good flight"

Amelia Earhart did not think of herself as a symbol or leader of a feminist crusade, but she was both. According to her friend Louise Thaden, her personal ambitions "were secondary to an insatiable desire to get women into the air, and once in the air to have the recognition she felt they deserved." In lectures, interviews and in print she was outspoken in her condemnation of an educational system that "goes on dividing people according to their sex, and putting them in little feminine or masculine pigeonholes." She was equally critical of a society in which "girls are shielded and sometimes helped so much that they lose initiative and begin to believe the signs, 'Girls don't,' and 'Girls can't,' which mark their paths."

To other women, Amelia Earhart was a revelation and an inspiration. She was not particularly pretty by conventional standards, but she was utterly feminine and devastatingly attractive in an unstudied way. She was intelligent without being overbearing, gently spoken yet persuasive, quick to smile and make a joke, successful, modest, self-reliant—all the things that many women dreamed of being.

People would ask her, or she would feel obliged to explain, what had motivated her to undertake her dangerous long-distance flights. "Women must try to do things, as men have tried," she wrote to her husband, George Putnam, before one particularly dangerous venture. "When they fail, their failure must be but a challenge to others." But she had prefaced that remark with an almost childlike, "I want to do it because I want to do it."

Simply wanting to do something struck some critics as a trivial reason for endangering one's life. But AE, as she liked to be called, believed that it was a perfectly legitimate motive, a way of being true to herself. Above all else, Amelia Earhart was true to herself, and she continued to pour her considerable energies into her aviation career.

On August 24, 1932, three months after her transatlantic solo, she flew her Lockheed Vega nonstop from Los Angeles to Newark, New Jersey, covering a distance of 2,478 miles in 19 hours and 5 minutes, and setting a new long-distance record for women. The flight time was a personal record as well, marking the longest time AE had yet spent in

Soaring over the Bay Bridge as she sets off from Oakland, California, in 1937, Amelia Earhart embarks on the first of two attempts to fly around the world in her specially equipped Lockheed Electra.

This air marker at Selkirk, New York, gives a pilot directions to the nearest airfield, half a mile away, and to Albany, 15 miles away.

Air markers: the road signs of the skies

Noting that "an air route without markings is like a highway without signs," Phoebe F. Omlie of the National Advisory Committee for Aeronautics in 1935 conceived a plan to paint town names and directional indicators on the roofs of buildings throughout the United States.

At Eleanor Roosevelt's suggestion, the Bureau of Air Commerce hired women fliers, among them Louise Thaden and Blanche Noyes, to scout sites and get permission for the markers from local officials and building owners. Workmen then painted the signs in orange characters, seven to 12 feet high and legible from 3,000 feet. The system soon became the answer to a lost pilot's prayer, with 16,000 markers—one every 15 miles on every air route in the country.

the air alone. Then, slightly less than a year later, in July 1933, she repeated the coast-to-coast flight, setting a new women's speed record for the route by cutting her own previous time by two hours.

For all their novelty, the transcontinental flights lacked the dramatic appeal of oceanic solos, and they made relatively little impact on a public that was coming to expect greater things of the world's foremost woman flier. Yet the long overland flights were important preliminaries in AE's progress toward a far more ambitious goal—to become the first aviator, man or woman, to fly alone from Hawaii to California, crossing a blank and challenging stretch of some 2,400 ocean miles.

To prepare for the transpacific flight, she and her husband, GP to her AE, moved in 1934 from their home in Rye, New York, to the West Coast, where they would be closer to the heart of the aviation industry. AE sold her old Lockheed Vega to the Franklin Institute in Philadelphia and bought a new, improved model with auxiliary navigational equipment. To her regret, her old colleague Bernt Balchen was not on hand to supervise preparations for the flight, but AE found in Paul Mantz an excellent replacement.

At 31, Mantz was a professional Hollywood daredevil who operated his own small, specialized fleet of planes and flew stunts for such motion pictures as *Hell's Angels* and *Men With Wings.* Flamboyant as he was, his own life depended as much on his technical proficiency as on his flying skills, and he was the ideal technical adviser, engineering consultant and navigation instructor.

Late in December 1934, Amelia Earhart, Putnam and Mantz sailed from Los Angeles to Honolulu on the Matson Line's S.S. *Lurline,* with AE's Vega lashed to the deck. They had tried to keep their plans secret, but reporters who met the ship were naturally intrigued by the Earhart entourage, complete with plane, and were curious about AE's intentions. Fueling their curiosity was the news that a group of Hawaiian businessmen, seeking to publicize their island paradise, was offering a $10,000 purse for a flight from Hawaii to California.

It was a welcome prize. Record flying was an expensive business; an opportunity to make a self-supporting flight did not arise very often. One of the lures of the Pacific venture was that the expenses of a successful flight would be covered by the Hawaiian sponsors.

Ironically, this attractive financial factor almost sabotaged the venture. As Amelia Earhart waited patiently for Mantz to check out the new plane and for favorable weather conditions along her projected course, business rivals of the sponsoring group charged that the purpose of the enterprise was to wangle tariff concessions for Hawaiian sugar. Mainland journals inflated the accusation by suggesting that the famed pilot was being manipulated by the sugar interests, who were using her prestige to persuade Congress to reduce the tariff on sugar shipped from the Territory of Hawaii to the United States.

Alarmed by the uproar, the flight's sponsors had second thoughts about proceeding and summoned Amelia Earhart from nearby Wheeler Field to an emergency meeting in the private dining room of the Royal Hawaiian Hotel. She listened, thunderstruck, as they proposed that she give up the venture and ship her plane back to the mainland.

The resolute Amelia Earhart would have nothing to do with such a suggestion and coldly accused her intimidated backers of cowardice. They knew, she said, that the stories about her alleged political influence were totally false, yet they seemed ready to give in to their rumor-mongering rivals. As for herself, she was firmly committed to her plans. "I intend to fly to California within this next week," she said, "with or without your support." With that, she left the room. Her sponsors, perhaps shamed by her example, reconsidered their position and decided to continue their support.

Four days later, on January 11, 1935, Amelia Earhart was ready for takeoff. But the weather seemed to conspire against her. Early that morning the ordinarily mild and sunny skies of Hawaii began to fill with clouds driven by a brisk southwesterly wind. By 11 a.m. a tropical downpour was drenching Honolulu, including the unpaved runway of Wheeler Field. Watching the relentless rain as the hours went by, she knew that if she left that afternoon, as planned, she would have to lift her heavily loaded plane from a sea of mud.

When the rain slackened at about 3:30 she checked the field and found it as slushy as she had feared. Takeoff would now be messy and difficult, perhaps impossible. Studying the weather forecast, she saw that if she did not leave that afternoon she might be held up for 10 days or more by summer storms brewing up over the central Pacific; but if she could get off before a new cycle of bad weather moved in, she had a good chance of hitting fair weather along her projected route.

So far as AE was concerned, it was now or never. By 4:30 the Vega was out of its hangar, warming up on the concrete apron. She settled into the cockpit and taxied to the head of the runway. As she paused before her takeoff run the Vega's wheels sank into two or three inches of sticky mud. With 500 gallons of fuel, a two-way radio, a rubber raft and several other pieces of emergency equipment, the plane weighed at least three tons and was going to be a monster to lift.

Out of the corner of her eye Amelia Earhart noticed three fire engines and an ambulance drawn up near the hangars. Every man in a detachment of soldiers was armed with a portable fire extinguisher. And every woman present, AE noted, seemed to be clutching a handkerchief, ready for any emergency.

Oddly cheered by this apparent pessimism, she opened the throttle and released the brakes. The Vega lumbered along reluctantly, straining against the pull of the mud and its own load. In the distance AE could see mountain peaks shrouded in low-lying clouds; much nearer, down the runway, stood a pair of checkered flags marking the point where she would have to cut the throttle and slam on the brakes if she had not yet become airborne. Paul Mantz dashed alongside the plane, shouting: "Get that tail up—get that tail up!" Then he was left far behind. She felt

A 1935 cartoon in the Pittsburgh Post-Gazette depicts Amelia Earhart tomboyishly leaping over seven famous male fliers as she flawlessly performs a "new stunt," the first solo flight ever made from Hawaii to the continental United States.

the ship getting gradually lighter. The tail rose. The wheels hit a bump on the runway and the Vega jolted into the air. She pushed the throttle full ahead and held it there; the plane, starting to sink back, caught the new surge of power and lifted off. With a matching surge of exultation, AE took the Vega up to 5,000 feet, banked over Diamond Head and climbed through the cloud layer into a clear blue evening.

The hardest part was over. The rest of the flight was so smooth that it would even have been something of an anticlimax had it not been such a triumph. Stars lit the balmy night sky, and a blazing sun brought the dawn. All instruments, including the radio, functioned perfectly, and AE set the Vega down at Oakland just before noon on the day after takeoff. She had flown without rest for 18 hours and 15 minutes, and California was waiting for her.

As she taxied toward the ramp at the end of the runway she saw a mass of people—the crowd was estimated at 10,000—waving and cheering. Barriers toppled, and the crowd swarmed to meet her. She cut the engine, locked her brakes, stepped stiffly out of the cockpit and grinned cheerfully at her welcomers. Amelia Earhart had added one more record flight to her growing list.

Three months after her Pacific flight AE made a 1,700-mile air journey from Burbank, near Los Angeles, to Mexico City. Although she actually stopped once—in the desert just 50 miles north of Mexico City, to remove a bug from her eye—she did not refuel and the flight was officially considered to be nonstop, another first for any pilot. But this record-making trip was not simply an end in itself; it was also a way of getting to the starting point of an even more significant flight.

Before leaving for Mexico Amelia Earhart had discussed her plans with her friend Wiley Post, the well-known pilot who had made two round-the-world flights. She would fly, she said, from Burbank to Mexico City, then from Mexico City to New York. On the latter flight she would take the most direct route—specifically, across the Gulf of Mexico and over New Orleans. Post's warnings about the hazards of the long overwater flight may have made her all the more determined to try it, since she suspected that Post really meant to say that the venture was too dangerous for a lone woman pilot.

Amelia Earhart arrived in Mexico City on April 20, 1935. After a round of festivities and the construction of a makeshift, extra-long airstrip to accommodate her heavily loaded Vega's takeoff run, she was ready to return to the United States.

Early in the morning of May 8, taking advantage of the first favorable weather forecast in more than a week, AE guided the Vega down the runway until the plane built up a speed of more than 100 miles per hour. Then, as the flier later recounted, the Vega "just flew itself into the air."

She looked down upon a fairyland of majestic mountains and tiny adobe towns; then she came to the shimmering waters of the Gulf dappled with fleecy clouds. Fourteen hours and 19 minutes after takeoff she was greeted at Newark Airport, in New Jersey, with almost hysterical adulation. It took a squad of burly policemen to rescue her from a screaming mob of well-meaning admirers.

Following her record-setting Mexican flight, Amelia Earhart decided to take up a less rigorous life for a time; in early June, she accepted an appointment as visiting aeronautics adviser at Indiana's Purdue University, which was building up a comprehensive aviation department. Before settling into the academic routine, however, she found the time and the inclination for another aerial venture, the 1935 Bendix race from Los Angeles to Cleveland. And in the course of this annual event she would come to know the tough and ambitious Jacqueline Cochran, a rising star among female fliers and one of many women whose interest in flying had been sparked in part by the example of Amelia Earhart.

By her own account of her life, in which some of the early details are obscure, Jackie Cochran was born amid the squalid sawmill camps of northern Florida, probably in 1912, and was orphaned as an infant. When she was eight years old, she moved with her foster family to Columbus, Georgia, where she went to work in a cotton mill that paid all of six cents an hour for a 12-hour shift. Living in poverty, she nurtured dreams that she would one day have fine clothes and a car and would see the world. Thrown out of work when a labor dispute shut down the mill, young Jackie left her foster family and hired herself out as live-in help for the owner of a beauty shop. It was hard work, but she soon learned the beautician's trade, mastering the arts of coloring and styling hair and applying the newly introduced permanent wave.

After saving several hundred dollars, she left Columbus and went to work in a Montgomery, Alabama, beauty shop. Her first and best customer, a juvenile-court judge, took an interest in her and persuaded her to go into nurses' training to help others. Jackie Cochran studied three years in an Alabama hospital, but a few weeks of working with a doctor in her native Florida lumber country convinced her that she could do virtually nothing to alleviate the sufferings or brighten the lives of the people there. She determined that, if she was ever going to do anything for herself or anyone else, she had to get away and make money.

Dipping into what was left of her savings, Jackie Cochran became a partner in a Pensacola beauty shop. Then, not yet in her twenties, and at a time when jobs were hard to come by, she struck out for New York and coolly landed a job with the exclusive Antoine's Salon at Saks Fifth Avenue. Skilled, popular with the customers and completely self-possessed, she was soon commuting between the New York salon and Antoine's Miami Beach shop. Miami socialites welcomed her into their homes and clubs and gambling casinos, but she still felt confined by her routine work. And then, at a Miami dinner party early in 1932, she met Floyd B. Odlum, a Wall Street lawyer and investment genius who had made his first few million dollars several years earlier by the age of 36.

Odlum, tired of the usual party chitchat, was intrigued with the pretty young woman who sat next to him and spoke so animatedly about her

After her 1935 Hawaii-Oakland flight, Amelia Earhart and her husband, George Putnam (right), grant an interview to newsmen beside her Lockheed Vega at the Newark, New Jersey, airport. Putnam, a New York book publisher, was an aggressive publicity seeker on his wife's behalf.

dream of escaping Antoine's plush salons by going on the road for a cosmetics company. Odlum observed that she would make a most effective sales representative for any business, but he also pointed out that competition was so keen because of the Depression that she would almost need wings to cover her territory fast enough to hold her own.

Back in New York, Jackie Cochran gave a great deal of thought to this casual comment. If she could pilot a plane, she reasoned, she could beat out any other salesperson on the road. In the summer of 1932, while Amelia Earhart was breaking one long-distance record after another, Jackie Cochran took a three-week vacation to learn to fly at Roosevelt Field, on Long Island. "When I paid for my first lesson," she wrote afterward, "a beauty operator ceased to exist and an aviator was born." She got her pilot's license before the three weeks were up, and that winter she drove cross-country and entered the Ryan Flying School at San Diego to get more flying experience.

At the urging of a Navy pilot friend from her Pensacola days she soon bought her own plane—an old Travel Air—for $1,200. The friend and some of his associates then gave her an intensive flight training course, and by 1933 she had earned her commerical pilot's license. The beauty operator was left far behind; now the businesswoman pursued her idea of selling cosmetics, said good-by forever to Antoine's and within two years had established Jacqueline Cochran Cosmetics, Inc., with a beauty salon in Chicago and a cosmetics laboratory in New Jersey. Her adviser and most enthusiastic supporter was Floyd Odlum, who became her ardent suitor and later, on May 11, 1936, her husband.

Almost as soon as she got the feel of controlling an airplane, Jackie Cochran realized that flying meant more to her than just transportation for sales trips. She wanted to race. Her first major competitive effort was the 1934 MacRobertson Race from London to Melbourne, Australia, which offered big prize money and an opportunity to see much of the world. Unfortunately, her attempt was dogged by mechanical problems. Her first race plane, a long-range Northrop Gamma, crash-landed on its delivery flight. Its replacement, a specially designed Gee Bee racer, led the field as far as Bucharest, Rumania, where it broke down and had to be grounded. The disappointed pilot sent her plane back home by ship and sold it.

Meanwhile, the Gamma had been rebuilt just in time for the annual Los Angeles-to-Cleveland Bendix race in the late summer of 1935. Once again, however, the craft was plagued by malfunctions. Representatives of both the plane maker and the engine builder urged Jackie Cochran to spare them from an embarrassing mishap by dropping out of the competition. To make matters worse, a heavy fog began rolling in shortly before the race was to start. Knowing that her withdrawal would be chalked up to a womanly fear of foul-weather flying, she was determined to stay in the race.

Her resolve may well have been stiffened by Amelia Earhart, the only other woman in the contest, whom she had just recently met for the first time. About a half an hour past midnight on August 31, the fog cleared briefly and AE lifted her Lockheed Vega off the runway. Behind her in the compact passenger cabin, Paul Mantz and a friend were calmly pouring drinks for themselves and setting up a game of gin rummy. Both pilot and passengers knew that the sturdy Vega was outclassed and had no chance of winning, but they hoped to at least place and win enough money to cover expenses.

After the Vega's takeoff, the fog closed in again but at about 2 a.m. lifted sufficiently for the start to resume. One by one the other contestants flew off into the night. Then at 4:22 a.m., Jackie Cochran started down the runway in her temperamental Gamma. But the engine was not producing the power she needed, and by the time she was halfway down the runway she doubted that she would ever get into the air. Praying fervently, she forced the engine to give everything it had; just as time and yardage ran out she felt the plane's wheels leave the ground.

Once aloft, the craft felt unresponsive and sluggish. Its air speed was so low that it was some time before the uneasy but determined pilot could turn it on course and reach cruising altitude. By the time the straining Gamma broke out of the bad weather over the Arizona desert, the plane was shuddering and its engine was badly overheating. The Grand Canyon loomed ahead, and beyond it a violent electrical storm was brewing. She decided that enough was enough. Shortly after daybreak she turned back to Kingman, Arizona.

Amelia Earhart's "bare essentials"

Unlike some female aviators who loaded their planes for marathon flights with food for the trip and fancy dresses to wear on arrival, Amelia Earhart took a certain ascetic pride in traveling light. Her only sustenance for her 15-hour-18-minute solo flight across the Atlantic in 1932 was a thermos of soup and a can of tomato juice. "A pilot whose plane falls into the Atlantic is not consoled by caviar sandwiches," she insisted. "Everything but the bare essentials would have distracted my attention."

Her selection of items was a revealing exhibition of her optimism, pragmatism and modesty. Some of her "bare essentials"—such as tomato juice, scarves and a lucky bracelet—actually seemed barely essential, but her emergency gear was more sensible, consisting of the few light tools needed to escape a downed plane and signal for help.

Although she cared about her appearance, she never packed a dress, since rich admirers at her destinations were pleased to welcome the famous flier with the use of a wardrobe full of fancy frocks.

Amelia Earhart's powder compact was a priority accessory on all journeys. It helped her "look nice when the reporters come."

This elephant's-foot bracelet with silver inlay brought her good luck, she believed. But she left it behind on her last flight.

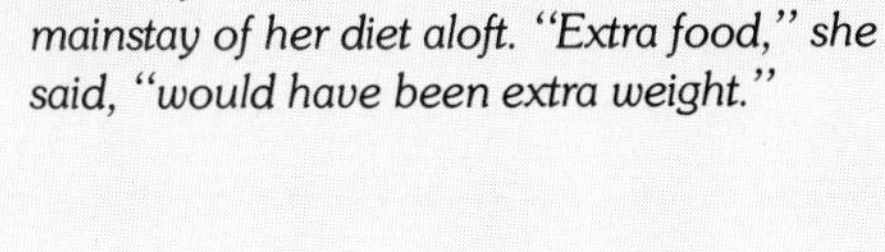

Tomato juice in cans like this one was a mainstay of her diet aloft. "Extra food," she said, "would have been extra weight."

A gaily colored silk scarf was her "single elegance" on the 1928 Friendship flight, and afterward it was one of her trademarks.

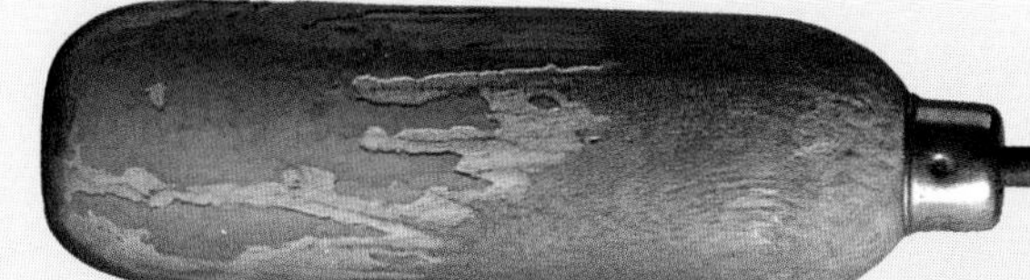

The sharpened point of this ice pick was handy for punching small holes in the tomato juice cans (above) she sipped from.

A bottle of smelling salts kept Amelia Earhart awake and alert during long flights. She did not drink coffee or tea.

If the aviator was forced down at sea, her emergency light, which could be ignited in water, would send a signal for help.

"I dumped gas for a landing," Jackie Cochran wrote later, "and the gas came back into the cockpit and drenched me from head to foot. I went ahead and landed, assuming that if I got a spark anywhere, even from a rock hitting the plane, I would catch on fire. I taxied up, piled out of the airplane, ran to the ladies' room, stripped off my clothes and started to wash."

Some hours later, Amelia Earhart and her two card-playing passengers landed sedately in Cleveland just two minutes before the deadline. They had finished in fifth place, winning $500 in prize money.

Soon after the 1935 Bendix run, Jackie Cochran began to concentrate more on her cosmetics business, but at the same time she laid plans to make better showings in future air races. Amelia Earhart, meanwhile, took up her academic duties at Purdue, where she moved into a women's dormitory room in order to make herself more accessible to students. But even as she counseled her charges on woman's role in aviation and in society, she was planning for yet another great personal achievement. Having crossed the Atlantic, having bridged the Gulf of Mexico and flown the Pacific from Hawaii to California, she was now determined to circle the globe by air.

Amelia Earhart had begun to think seriously about an around-the-world flight during her crossing of the Gulf of Mexico in May of 1935. It had occurred to her then that such a flight would best be made in a well-equipped twin-engined plane, but such a craft had seemed out of her grasp until she had settled in at Purdue. Then the university authorities, impressed by her performance, established the Amelia Earhart Research Foundation and started a fund for the purchase and maintenance of an aircraft that she could use as a flying laboratory. A number of aviation-related corporations gave to the fund, and AE soon had the plane of her choice: a sleek twin-engined Lockheed Electra with a top speed of more than 200 miles per hour and plenty of space—it normally accommodated 10 passengers—for extra equipment and fuel tanks.

The new Electra was intended primarily for research flights to test such things as the effects of altitude on metabolism, and the rate at which pilot fatigue would be brought on by the concentration demanded by complex aircraft instrumentation. AE was also interested in other questions. "Are men and women different in their reaction to air travel?" she wrote. "If so, how? And perhaps, why?" At the same time, though, she knew that it might be possible to combine such solid experimental efforts with the epic flight that she wanted to undertake simply because she wanted to do it. The days of adventure seemed almost gone from the earth, but there was one enterprise left for a trailblazing flier: a globe-girdling flight at the very bulge of the Equator. It was true that several round-the-world flights had already been made. But none of the pilots had taken on the earth's full circumference of some 25,000 miles—they had followed instead a shorter route north of the Equator. What AE wanted to do was to circle the earth near its midsection, thus becoming the first woman to fly around the world and the first flier of either sex to take the longest, toughest course (her planned route, with its necessary zigzags between airfields, was some 29,000 miles long).

Paul Mantz was once again called in as technical adviser, and the Electra's fuselage, which had been fitted with extra fuel tanks that would give the plane a cruising range of 4,000 miles, was equipped with a modern navigation compartment *(pages 130-131)*. For several weeks after she took custody of the plane in July 1936, AE "wore a groove between San Francisco and Southern California," as she put it, getting used to her new ship and working the kinks out of it. Not quite all the kinks were out by September, however, when she and her copilot—Helen Richey, who had flown commercial airliners until strictures imposed by the all-male pilot's union compelled her to quit—flew the plane from New York to Los Angeles in the 1936 Bendix race. Trouble developed in the Electra's fuel lines, and the navigation hatch fitted so badly that it blew open and caused trouble for half the flight. Amelia Earhart again finished in fifth place—her disappointment tempered by the knowledge that another woman pilot, Louise Thaden, had won the usually male-dominated contest *(page 120)*.

While the Electra went back to the shop for more work, Putnam made arrangements to have spare parts, fuel and oil supplies cached along AE's intended round-the-world route and patiently obtained landing clearances and other official documents from the countries to be visited or flown over. AE studied maps and charts, weather patterns, instrument flying and emergency landing fields. Reluctantly, she decided that she would need a navigator for the long Pacific stretches.

Putnam believed that an ideal candidate for the job would be Henry Bradford Washburn Jr., the 26-year-old head of Harvard University's Institute of Geographical Exploration. Expedition leader, geographer, flier and aerial photographer, Washburn had first come to GP's attention almost 10 years before as a teen-age Alpinist who wrote an article about his experiences. He was now teaching field astronomy, the use of celestial bodies for extremely precise ground navigation, at Harvard. AE, cautious about selecting a partner for such an undertaking, agreed to meet the young explorer and discuss the projected flight—but without committing herself to offering him the job.

Washburn was invited to the Putnam home in Rye. There, sitting on the living-room floor, he followed the track of the famed flier's finger as it traced her proposed itinerary on the maps spread out in front of them. "We started around the world just literally from airport to airport where the flight was going to go," he recalled later. The route ran from Oakland to Honolulu; thence to tiny Howland Island in the central Pacific, a critical refueling stop; from Howland eastward to New Guinea; then to Darwin, Australia; and eventually across Africa and the South Atlantic to Brazil and back to the United States.

Washburn stared at the expanse of the Pacific and at Howland Island, less than two miles long and one-half mile wide and far from any other

With a 420-hp engine and a light, largely wooden airframe, Louise Thaden's Staggerwing achieved a top speed of 210 mph—faster than many 1930s Army fighters. Retractable landing gear, unusual then on private planes, and a sloping windshield reduced drag, as did the tapered wings with their smooth connecting I strut.

A beautiful blue Bendix winner

Not only the public but other fliers as well were astonished when Louise Thaden and her copilot Blanche Noyes won the coveted Bendix Trophy in 1936—not because the winners were women but because they had captured the trophy and $7,000 in prizes in a small stock biplane rather than a souped-up racer or a large twin-engined machine made for such long transcontinental hauls.

The plane Louise Thaden flew to victory was no ordinary aircraft, however, but a sleek triumph of aeronautical design produced by airplane maker Walter Beech, the Beechcraft Model 17. It was called the Staggerwing because its lower wing was set forward of the upper wing, unlike most other biplanes, which had the lower wing set to the rear of the upper. This unusual arrangement gave the pilot excellent upward visibility and lent the craft a memorable, forward-thrusting grace; as one aerohistorian wrote, Staggerwings were "among the most enduringly beautiful airplanes ever made."

Flanked by the famous trophy and its donor, Vincent Bendix, Louise Thaden gives her victory speech at Mines Field in Los Angeles. Her flight took 14 hours 55 minutes.

landfall. According to Washburn, he then asked the flier how she was going to find Howland Island. "And she said, 'Well, we're going to do it by dead reckoning.' I said, 'Holy smoke! That's 2,000 miles. Howland Island is just a tiny needle in an enormous haystack of water out there, and you've really got to have a radio there to do this.' " He suggested a transmitter that would emit a constant signal from the island and a direction finder in the plane that would enable her to home in on the signal. Amelia Earhart did not think such a precaution was necessary; Washburn, still insisting that it was, went upstairs to bed.

In the morning he packed his bag and left. He had not been asked to navigate and later indicated that if he had been, he would have declined. For all the experience that he had crammed into the previous 10 years, he did not feel qualified to pinpoint tiny Howland Island without a homing signal.

His arguments apparently had some effect, however, because before the flight began arrangements had been made to have a Coast Guard cutter standing off Howland Island with a transmitter that could send a homing signal, and the Electra was fitted with a direction finder. On the afternoon of March 17 1937, AE took off from Oakland with a crew consisting of adviser Paul Mantz and navigators Harry Manning and Fred Noonan. Mantz, making a final check of the Electra as copilot on the first lap of the flight, would remain in Honolulu. Manning, a sea captain on leave from the American President Lines, was to serve as chief navigator as far as Australia, with veteran flier Noonan assisting him on the Honolulu-to-Howland flight. With the Pacific stretch behind her, Amelia Earhart would fly on alone.

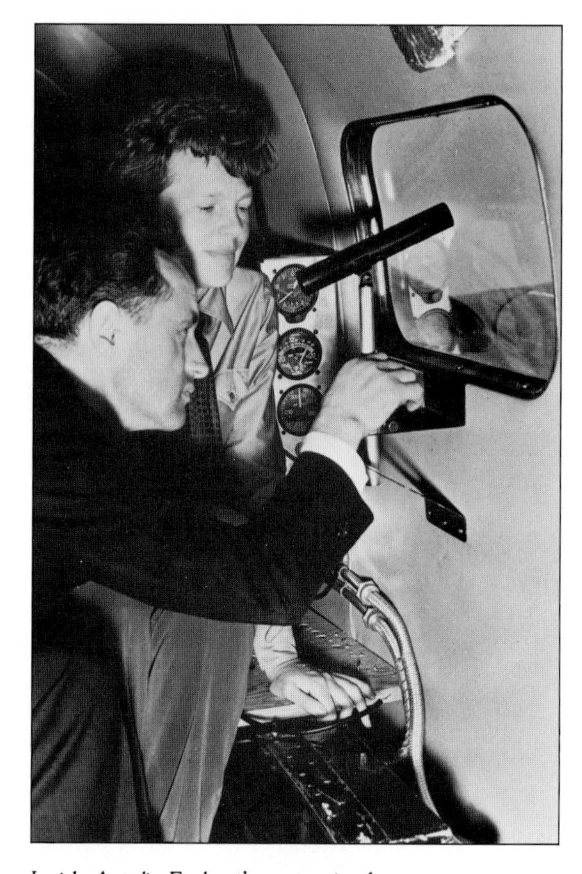

Inside Amelia Earhart's customized Electra, navigator Harry Manning practices taking celestial bearings through the scope of a navigational instrument called a pelorus while Amelia Earhart looks on. The instrument gauges visible behind her on the cabin wall measure altitude, air temperature and air speed.

The Oakland-to-Honolulu flight was uneventful and took less than 16 hours. Then on the morning of March 19, after a day's weather delay, the Electra ground-looped during its takeoff run for the Howland Island hop. The pilot and her two navigators escaped unhurt, but the plane was badly damaged. Its wheels were sheared off and the undercarriage was smashed; the right wing was crumpled, the right motor mount was bent and all the propeller blades were twisted.

The Electra was shipped back to the Lockheed factory in Burbank, where it was rebuilt and structurally strengthened within two months. But the delay in flight plans had changed many things. Harry Manning's leave of absence had run out and he had returned to his shipboard duties, leaving Fred Noonan to serve as the sole navigator. A thoroughly experienced transport pilot and aerial navigator, Noonan was an ideal flying partner. He had worked for Pan American Airways for many years and had helped plot the airline's transpacific routes by way of Midway, Wake, Guam and the Philippines. Pan Am had discharged him because of a drinking habit that he gave up when he began his association with Amelia Earhart, who had implicit faith in the cool and unflappable aviator.

She was in sore need of such a trustworthy companion, for she had decided not to fly solo on any part of the trip. Worldwide weather

The Lockheed Electra, its landing gear crushed and right wing damaged, lies crippled on the runway at Pearl Harbor after crashing during an attempted takeoff for Howland Island. Atop the plane (left to right) technical adviser Paul Mantz, Amelia Earhart and conavigator Fred Noonan investigate the mishap.

conditions were changing with the seasons, and the delayed departure date made it preferable to fly east rather than west. Instead of heading from Oakland to Honolulu and then to minuscule Howland Island on the first stage of the flight, Amelia Earhart would start out across the Atlantic and leave the Pacific stretch for last: from New Guinea to Howland, and from Howland to Hawaii and then home. She would now be needing her navigator more at the end of the flight than at the beginning.

The reversal in the route also demanded a whole new set of arrangements. AE and her team had to study new charts, arrange for fuel, oil and spare parts to be shifted to new locations, alert different landing fields and reassign mechanics, unravel more international red tape and obtain new permissions. As originally planned, the United States Coast Guard would have a cutter standing off Howland Island so that AE could talk to it and home in on its radio signals.

The Electra's own radio gear was also scrutinized afresh. Mantz had fitted the plane with the best equipment available commercially at the time: a transmitter and receiver designed for operation on the standard aviation frequencies as well as on the long-range, 500-kilocycle international emergency wavelength—which for best results required a trailing antenna wire at least 250 feet long. The signals sent out through this long antenna could be picked up by a direction finder aboard the U.S.S. *Itasca,* the Coast Guard cutter assigned to take station off Howland

Island. Thus while the aviators were using their own direction finder to pinpoint the island, the Coast Guard could get a fix on the plane's position. Also, Mantz had persuaded some Army Air Corps friends in Hawaii to provide an experimental high-frequency direction finder, which was to be taken to Howland by the *Itasca* and operated ashore. It could be a valuable adjunct to the *Itasca's* own direction finder.

On May 21, Amelia Earhart and Fred Noonan left Oakland in the rebuilt Electra for Miami. Ostensibly, it was a shakedown flight, but it was in fact the first leg of their eastward journey. For the next week a crew of Pan American mechanics made final adjustments to the plane's engines and instruments while AE waited with apparent calm, getting as grimy as a grease monkey whenever an extra hand was needed. Noonan, watching her, felt no qualms about the forthcoming flight. "Amelia is a grand person for such a trip," he wrote to his wife. "She is the only woman flier I would care to make such an expedition with. Because in addition to being a fine companion and pilot, she can take hardship as well as a man—and work like one."

The press had not been informed of the start of the flight, but AE talked freely to Carl Allen of the *New York Herald Tribune,* who had been chosen to write the story from accounts she would send back from her various stops. After this flight, she told him, she intended to give up long-distance flying and leave the spectacular ventures for a new generation of fliers. "I have a feeling there is just about one more good flight left in my system," she said reflectively, "and I hope this trip is it."

This map shows the route plotted by Amelia Earhart for her 1937 attempt to fly around the world from west to east. Starting from Oakland, California, she planned refueling stops at the points marked by dots. Squares indicate places where the aviator made unscheduled landings in the course of her flight.

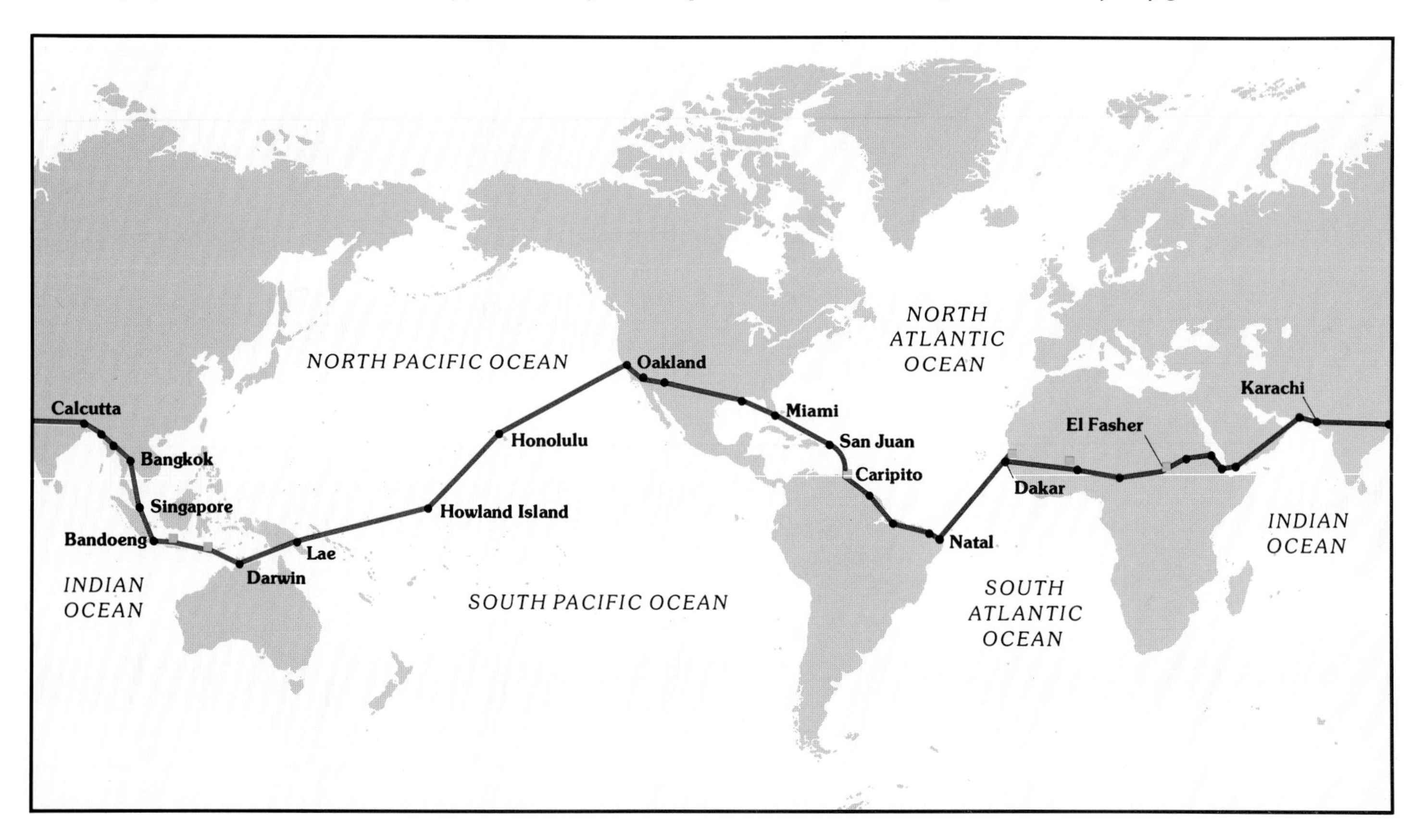

In the early-morning stillness of June 1, 1937, Amelia Earhart and Fred Noonan climbed into the waiting Electra at Miami's Municipal Airport. George Putnam waved good-by and watched the plane taxi to the takeoff runway, gather speed and sail into a cloudless sky, bound for Puerto Rico. After AE's arrival there, Putnam spoke to her by telephone. He then wrote to Paul Mantz, who had not been able to take part in final preparations for the flight. Upset when he heard that the global flight was getting under way earlier than he had anticipated, Mantz dropped a note to Putnam to inquire whether all final equipment and procedure checks had been satisfactory. "Between ourselves," Putnam replied, "the radio gave unending trouble. As I understand it, it was finally decided by the technicians that the longer aerials were improper. One part of them just canceled out the other, so they shortened the aerials and apparently got the thing pretty well licked." In fact, Amelia Earhart had left the trailing antenna behind in Miami.

Mantz was appalled. By abandoning the 250-foot-long antenna, the fliers had severely reduced the effectiveness of an enormously important navigational aid that would have been especially useful in the mid-Pacific. But the early part of the around-the-world flight proceeded without serious incident. From Puerto Rico, AE and Noonan flew in easy stages down the South American coast to Natal, Brazil, their jumping-off place for the transatlantic leg of the journey *(map, page 124)*. They left Natal on the morning of June 7, and as they crossed the ocean Amelia Earhart jotted down observations to be sent back from each stop to Putnam and the *Herald Tribune*. "Just crossing equator. 6,000 feet," she wrote at one point. "Sun brilliant. Little lamb clouds below."

After reaching Dakar, on the west coast of Africa, the two travelers droned across the continent and over the Gulf of Aden to Arabia, with AE noting along the way: "We crossed stretches of country barren beyond words, a no-man's land of eternal want." In a kaleidoscope of changing scenes they flew on to Karachi, Calcutta and Rangoon. Then they were off to Bangkok, Singapore and Bandoeng; then to Darwin, Australia; and finally to Lae, New Guinea, the last stop before setting out for Howland Island. Along the way, the name "Amelia Earhart" in large red or white letters on waiting 50-gallon fuel drums had become a familiar sight at prearranged stopovers, and AE's radio call letters, KHAQQ, had become almost as familiar as her name.

"Twenty-two thousand miles have been covered so far," she wrote in her logbook after arriving in Lae on June 30. "There are 7,000 more to go." She and Noonan had been flying for 30 days since leaving Miami and were beginning to feel the strain. Yet the longest and most hazardous parts of the trip were still to come: 2,556 miles from Lae to Howland; 1,900 miles from Howland to Hawaii; 2,400 miles from Hawaii to the journey's end in Oakland.

The Electra had been performing superbly. In the course of the punishing trip it had suffered a number of minor mechanical problems, but

Amelia Earhart (top center) and Fred Noonan (reclining) lunch at a Venezuelan airport, one of 28 stops on their second bid to circle the world.

these were nothing out of the ordinary for a hard-working aircraft. Noonan had a problem with the chronometers, however. In order to take correct navigational readings, he had to set these precision timepieces accurately, but neither the Electra's radio nor the sensitive receivers at Lae airport could pick up the exact-time signals broadcast by the United States Navy and the Bureau of Standards. This was disturbing. "Howland is such a small spot in the Pacific that every aid to locating it must be available," AE observed in the notes she would send to Putnam.

It was July 2 in New Guinea and July 1 on Howland Island, across the international date line, when pilot and navigator climbed into the Electra on the runway at Lae. They had stripped the plane of everything that was not absolutely essential to the flight, including most of their personal belongings, and the ship's fuel tanks were loaded to capacity with 1,150 gallons of gasoline. Theoretically, this would be enough for 4,000 miles, but even a slight head wind would cut into that range.

Amelia Earhart had mixed feelings about her departure. There had never been time enough at the stopovers to do much exploring, and in

Dutch mechanics at Bandoeng, Java, refuel the Lockheed Electra from gasoline drums delivered in advance. "Always we found my usual calling cards," Amelia Earhart wrote, "fifty-gallon drums of gasoline, each with my name printed large upon it in white or red lettering."

Resting in Karachi, the aviator finds that riding a camel can be nearly as hazardous as piloting a plane: "It was a startling takeoff as we rose," she wrote, "reminiscent of the first symptoms of a flat spin. Camels should have shock absorbers."

her Lae dispatch she had noted: "I wish we could stay here peacefully for a time and see something of this strange land." But then, looking eastward over the Pacific, she added: "The whole width of the world has passed behind us—except this broad ocean. I shall be glad when we have the hazards of its navigation behind us."

At 10 o'clock that morning the Electra lumbered down 3,000 feet of jungle runway and lifted into the air a bare 50 yards short of a cliff edge overhanging the sea. Then it climbed serenely to a cruising altitude of 8,000 feet. Its estimated arrival time was 6:30 the following morning, Howland time—which meant roughly 18 hours aloft.

With word of the Electra's takeoff from Lae, Commander Warner K. Thompson and the crew of the *Itasca* began their vigil off Howland Island. One group of radiomen was assigned to operate the ship's low-frequency direction finder, which would be tuned to the 500-kilocycle emergency band. The high-frequency direction finder requested by Mantz was set up on the island, where Radioman Frank Cipriani would attempt to pinpoint AE's signals beamed on standard frequencies.

The plan called for AE to transmit her call signal and pertinent flight information twice an hour, at a quarter-to and a quarter-past. The *Itasca,* in turn, would call the Electra by voice signal on the hour and half-hour, giving weather information and homing signals, and would also tap out the Morse code letter *A* as a homing beam.

At 5:20 p.m., New Guinea time, Amelia Earhart reported to Lae that she was past the Solomon Islands and dead on course. Weather reports indicated that she was probably encountering strong head winds, but she mentioned no difficulty. She was then almost a third of the way to Howland, and in a few hours the *Itasca* would begin sending its scheduled signals. But for most of the night, nothing was heard from her.

At 2:45 a.m., Howland Island time, her voice came through heavy static that all but drowned it out. "KHAQQ . . . cloudy . . . weather cloudy," she was heard to say. But that was all.

The *Itasca* proceeded with its scheduled transmissions, and at 3:45 a.m. AE was heard again, her signal somewhat stronger this time, assuring the Coast Guard vessel that she would be tuned in on the agreed-upon radio frequency: "*Itasca* from Earhart, *Itasca.* Broadcast 3105 kilocycles on hour and half-hour. Broadcast 3105 kilocycles on hour and half-hour. Overcast."

An hour later the *Itasca* received another message from the Electra, but it was drowned out by static. Then at 6:15 a.m., only 15 minutes before Noonan's estimated time of their arrival at Howland, AE called in and calmly requested a radio bearing. She would whistle into the microphone, she said, so that the *Itasca* could get a fix on her position. On Howland, Radioman Cipriani tried to get a bearing with his direction finder, but he heard only a brief whistle that was almost indistinguishable from the usual harmonic whines and squeals of Pacific radio waves.

The *Itasca* asked for a longer transmission, preferably on the 500-kilocycle emergency wavelength. There was no reply.

By now the sun was up. The *Itasca* rolled gently in a low swell, and smoke from the ship's funnel rose straight into a clear sky. To the north and west there was a heavy cloud bank, suggesting that AE might still be picking her way through murky weather.

The Electra's projected arrival time at Howland came and went, and still there was no response to the *Itasca's* increasingly urgent signals. In the cutter's radio shack, there was a growing sense of unease: Something must be wrong with the aircraft's receiver. But at 6:45, her scheduled broadcast time, Amelia Earhart's voice suddenly came in loud and clear: "Please take bearing on us and report in half-hour. I will make noise in microphone. About 100 miles out."

Once more she was on the air too briefly for the *Itasca* to get a fix on her signal. The ship kept signaling, transmitting a homing signal almost continously. Still, there was no reply. Maybe the problem was with her transmitter; maybe, if she sent on 500-kilocycles with her long antenna, the *Itasca's* own direction finder would be able to take a bearing on the plane. The vessel's radiomen did not know that she had abandoned her lengthy trailing antenna at the start of her journey, rendering her emergency 500-kilocycle transmitter almost useless.

At 7:18 a.m. the *Itasca* broadcast to KHAQQ and asked for a change in frequency. "Cannot take bearing on 3105 very good. Please send on 500 or do you wish to take bearing on us? Go ahead, please." Again, there was no answer.

Finally, at 7:42 a.m., Amelia Earhart's voice came in at full volume and with absolute clarity. "KHAQQ calling *Itasca,*" she said. "We must be on you but cannot see you. But gas is running low. Have been unable to reach you by radio. We are flying at 1,000 feet." By now, the Electra should have been visible from the deck, but there was no sign of it, nor was there an answer to the *Itasca's* acknowledgment.

At 7:58 Amelia Earhart was heard again: "KHAQQ calling *Itasca.* We are circling but cannot see island, cannot hear you." Then she asked for a return transmission. Her signal was even louder than before but her voice, usually cool and almost monotonous when broadcasting, sounded high-pitched and strained. The men in the radio shack wanted desperately to help her, to get a bearing on her, to talk to her—but something was terribly wrong with communications. She was obviously close, but where?

The radioman sent out a feverish string of dot-dash homing signals. Suddenly she broke in, responding directly for the first time: "We are receiving your signals," she said and explained that she could not get a bearing on them. "Please take bearing on us and answer on 3105 kilocycles," she said.

She sent a series of dashes, hoping to maintain contact. The *Itasca* called back, telling her that they could not get a bearing on 3105 and asking that she signal on her emergency frequency. Then all sound from the Electra ceased. Desperately, the *Itasca* called Radioman Cipriani on Howland to find if he had been able to get a fix on the plane's position.

But Cipriani had not; his direction finder's batteries had weakened and were almost dead.

The *Itasca* kept sending, pleading for an answer. The Electra was two hours overdue, and the atmosphere in the radio room was growing almost unbearably tense. On the *Itasca's* deck, spotters strained for a glimpse of the wayward plane. A great plume of smoke poured out of the cutter's funnel, a homing beacon for anything within 30 miles. Other than that, and the cloud bank still crouching to the north and west, there was nothing but water and the searing blaze of the morning sun.

And then, at 8:45 a.m., the vanished voice came in again to give a cryptic position report. "We are on the line of position 157-337," Amelia Earhart said. Sounding almost frantic now, she concluded: "We are running north and south."

Then there was nothing but silence. The *Itasca* continued to call KHAQQ by voice and by key, begging Amelia Earhart to come in, al-

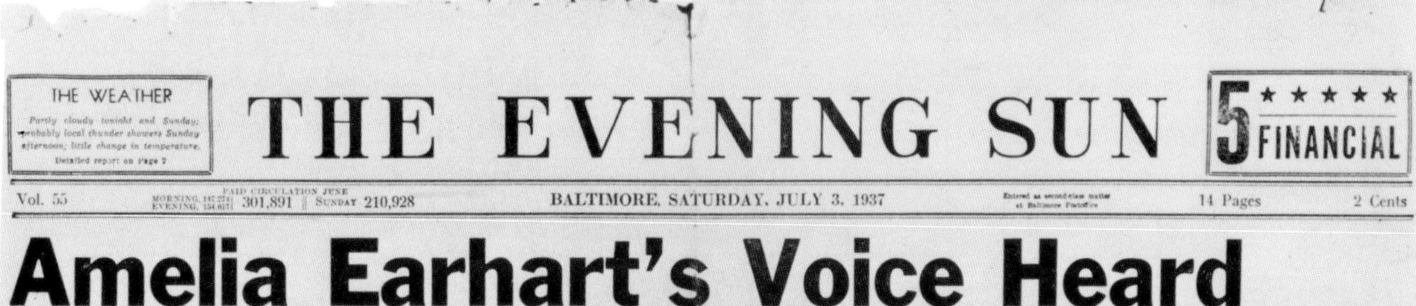

THE WEATHER

Partly cloudy tonight and Sunday; probably local thunder showers Sunday afternoon; little change in temperature.

Detailed report on Page 2

THE EVENING SUN

5 ★★★★★ FINANCIAL

Vol. 55 | PAID CIRCULATION JUNE 301,891 | SUNDAY 210,928 | BALTIMORE, SATURDAY, JULY 3, 1937 | Entered as second-class matter at Baltimore Postoffice | 14 Pages | 2 Cents

Amelia Earhart's Voice Heard Faintly By Radio Calling For Help

"So Big," And She Missed It

AMELIA EARHART

WEAK MESSAGE PLACES PLANE NEAR EQUATOR

Amateur Operator Recognizes Flyer's Voice. S O S Heard Earlier

BATTLESHIP JOINS WIDE PACIFIC SEARCH

Aviatrix And Aide Believed Down In Shark-Infested Waters

[By the Associated Press]

Los Angeles, July 3—Reports that the voice of Amelia Earhart had been picked up, calling "S O S" from the mystery spot where she is lost in mid-Pacific, buoyed hopes for her ultimate rescue today as the navy ordered a battleship into the search.

Two Los Angeles amateur radio operators said that as late as 7 A. M. Pacific time (10 A. M. E. S. T.) they distinctly heard her sound her c[illegible] letters, KHAQQ, after thrice sayi[illegible] "S O S" some twenty minutes earlie[illegible]

At San Francisco, however, a C[illegible] Guard sta[illegible] reported at noon [illegible] ern stand[illegible] time) it had recei[illegible] word wh[illegible] although radio [illegible] tion was [illegible]y good.

[illegible] were tun[illegible] quency of Miss Earhart's plane, [illegible] officials said "not a whisper" had bee[illegible] heard.

Earlier the Los Angeles operators, Walter McMenamy and Carl Pierson, interpreted radio signals as placing the plane adrift near the Equator between the Gilbert Islands and Howland Island, the latter Miss Earhart's

After Amelia Earhart disappeared over the Pacific, headlines like this announced that her SOS had been heard by two Los Angeles ham-radio operators. This news touched off a flurry of similar reports from short-wave buffs all over the United States. In the photograph she indicates how tiny Howland Island looks on the map.

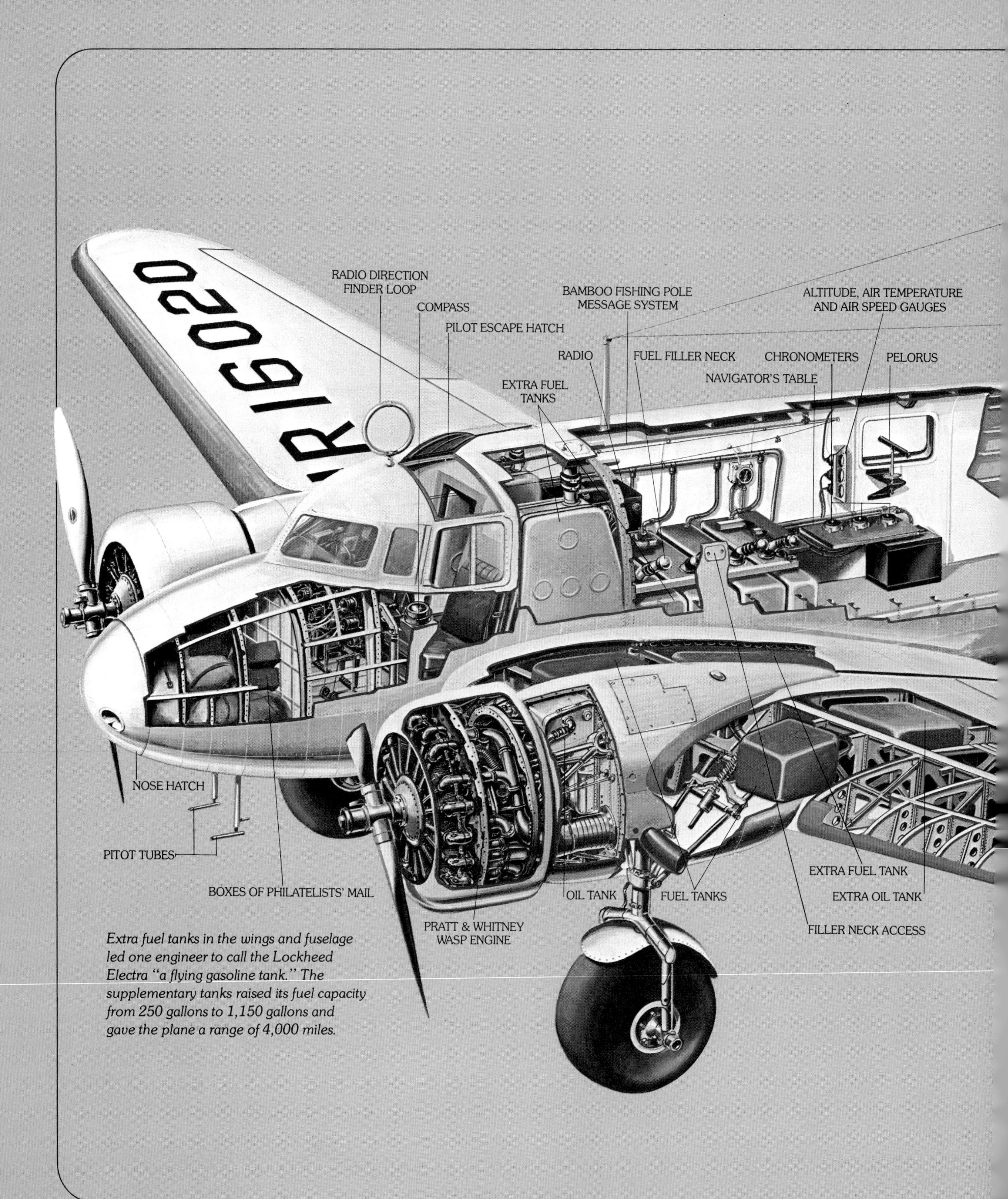

Extra fuel tanks in the wings and fuselage led one engineer to call the Lockheed Electra "a flying gasoline tank." The supplementary tanks raised its fuel capacity from 250 gallons to 1,150 gallons and gave the plane a range of 4,000 miles.

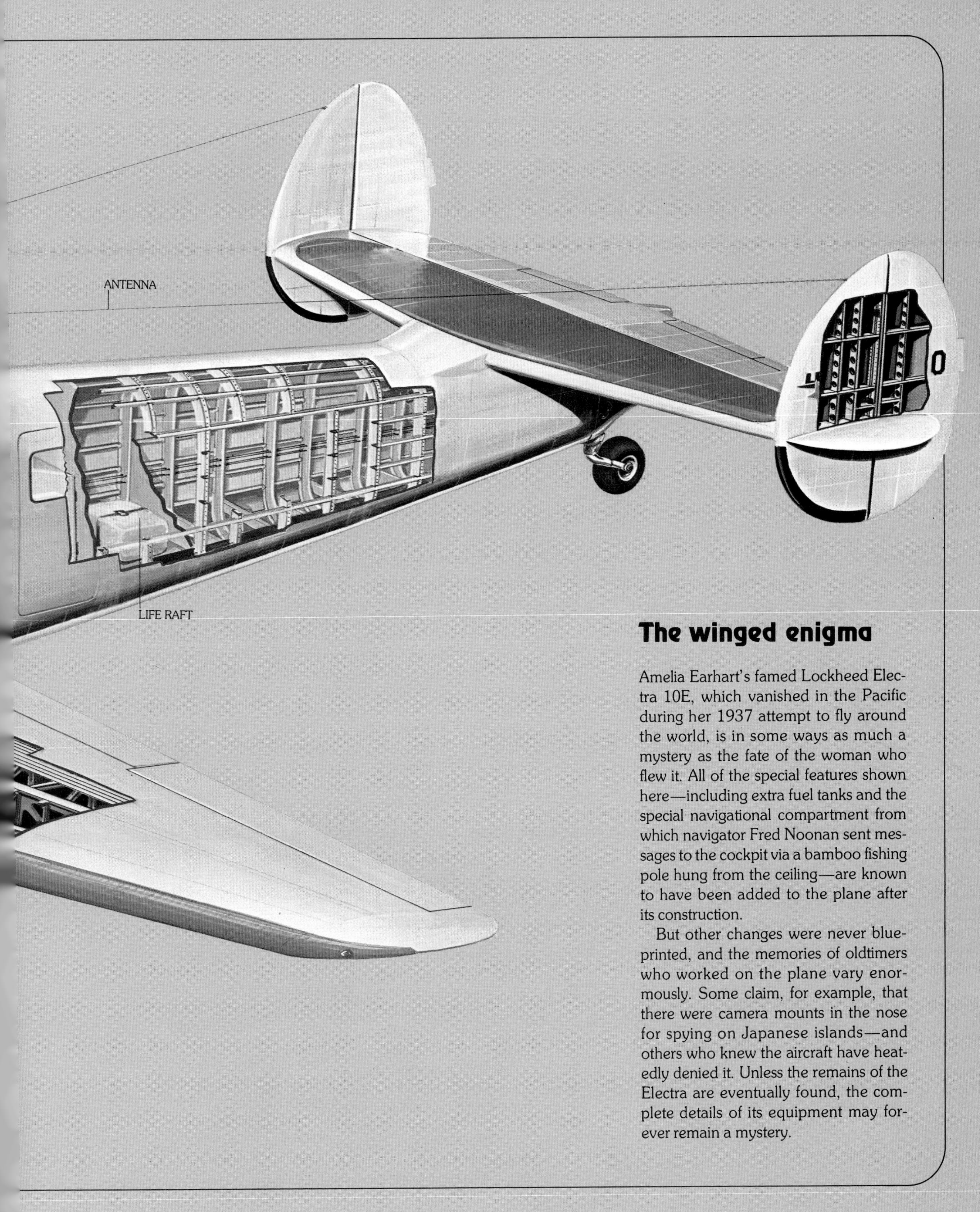

The winged enigma

Amelia Earhart's famed Lockheed Electra 10E, which vanished in the Pacific during her 1937 attempt to fly around the world, is in some ways as much a mystery as the fate of the woman who flew it. All of the special features shown here—including extra fuel tanks and the special navigational compartment from which navigator Fred Noonan sent messages to the cockpit via a bamboo fishing pole hung from the ceiling—are known to have been added to the plane after its construction.

But other changes were never blueprinted, and the memories of oldtimers who worked on the plane vary enormously. Some claim, for example, that there were camera mounts in the nose for spying on Japanese islands—and others who knew the aircraft have heatedly denied it. Unless the remains of the Electra are eventually found, the complete details of its equipment may forever remain a mystery.

though the men in the radio shack realized that she was probably gone.

Without any solid clues to guide them, the men of the *Itasca* began their search for Amelia Earhart at 10:40 a.m. But they had little idea where to begin. The plane's last position report had apparently been based on Noonan's observations of the sun, but since the Electra was "running north and south" it seemed that the two aviators did not know their exact location and were flying a search pattern in hopes of spotting Howland Island. Then there was the question of fuel: If AE's tanks were running low by 7:42, how much longer could she have gone on? Could she have ditched already? If not, surely she would do so soon.

The *Itasca* kept on signaling and searching, to no avail. As word of its mission reached the mainland, a sea search of unprecedented magnitude, involving some 10 ships and numerous planes, was launched to find the missing fliers. But they were not found. Days passed, then weeks, months and years.

A myth and a mystique grew around the disappearance. Conflicting rumors of the fliers' survival on one island or another, or of their execution by Japanese forces on Saipan or Truk, spread like wildfire after World War II, inspiring private searches and bizarre books spun out of manufactured mysteries. For all the lurid speculation, the real mystery remained. For a while she was close, and then she was gone. Exactly what happened, and why, and where Amelia Earhart and Fred Noonan went down would never be explained. But the achievements of America's best-known woman aviator would not be forgotten.

Besting a field of men in the 1938 Bendix race, Jackie Cochran climbs down from her Seversky AP-7 to claim $12,500 in prizes. Her victory led to a rumor that a man had secretly flown her plane for her.

Five months after Amelia Earhart's final message had crackled over the *Itasca's* radio, some 200 people gathered before a speaker's platform at Floyd Bennett Field on Long Island, New York. An icy November wind tugged at winter sleeves and made the American flags at both ends of the platform billow and snap. Yet speaker and audience paid little attention to the cold; they were there, as members of the Women's National Aeronautical Association, to honor the memory of Amelia Earhart. And in a way they were there, too, to mark the end of an era of pioneer women pilots. Jacqueline Cochran, who had become a close friend of the lost flier after meeting her in 1935, was the speaker.

"If her last flight was into eternity, one can mourn her loss but not regret her effort," she said with feeling. "Amelia did not lose, for her last flight was endless. Like in a relay race of progress, she had merely placed the torch in the hands of others to carry on to the next goal and from there on and forever." Indeed, Jackie Cochran would herself pick up that torch. And, in a world becoming jaded with daring flights having no immediately apparent purpose other than the fun of it, she would seem to be the first of a new generation of fliers, the prototype of a new breed of woman.

Following her failure to complete the Bendix race in 1935, Jackie Cochran had determined that her next competitive efforts would be more fruitful. She started winning in 1937, setting up one new speed record after another and then knocking it down herself. In the fall of that year she was the only woman in the Bendix contest; flying a Beechcraft Staggerwing against a field of race planes, she won $3,000 for coming in third and picked up a special purse of $2,500 for women's first place. Later, in December, she flew nonstop from New York to Miami in the spectacular time of 4 hours 12 minutes, breaking all previous records and paving the way for an even more dazzling achievement.

She had piloted a Seversky Executive on her record-shattering New York-Miami flight, and her stunning performance gave the plane's manufacturer an idea. Russian-born Alexander P. de Seversky had been trying for years to convince the United States Army Air Corps that his swift, long-range Seversky P-35 pursuit plane was the finest fighter in the world, but his pleas on behalf of his aircraft had engendered no interest. Now he thought that if Jackie Cochran, the woman flier with the suddenly luminous reputation as a speed queen, were to fly the civilian version of the P-35, the AP-7, nonstop from Los Angeles to Cleveland in the 1938 Bendix, someone would surely take notice. She jumped at the challenge. No pilot had yet tested the newest model of the 1,200-horsepower Seversky aircraft—an experimental plane with extra fuel tanks integrated into the wings—over long distances.

When race day came Jackie Cochran was the lone woman in a field of 10 entrants. She was the third to take off from Burbank airport, very early in the morning of September 3, and to Seversky's enormous gratification, and to her own, she was the first to land in Cleveland that afternoon. Despite a blockage in the AP-7's fuel lines, which had forced her to hold one wing higher than the other to feed the fuel to the engine, she had flown the 2,042 miles to Cleveland in 8 hours 10 minutes 31 seconds, completing the course nonstop and maintaining an average speed of a fraction under 250 miles per hour. And then, after touching down in Cleveland, she flew on to New York, adding a new women's west-east transcontinental record to her Bendix race victory. She had reached a high point in a competitive career that would be studded with records for first, best and fastest for years to come.

Recognition and trophies began pouring in. For three years running, starting in 1938, Jackie Cochran received the coveted Clifford Burke Harmon Trophy as the outstanding woman flier in the world. In 1939 she set a new women's altitude record, became the first woman to make a blind landing and set a new international women's speed record. In 1940 she set two speed records for men as well as women. And then, in June of 1941, she was the first woman to fly a warplane across the Atlantic.

It was an American bomber, and Jackie Cochran had picked it up in Canada for delivery to embattled Britain. For the world was once again at war; there were more important things for aviators to do than engage in friendly competition. And women fliers, who had struggled in vain for the right to serve in the air during the First World War, were determined to put their flying skills to practical use in the Second.

5

Women at war

On September 6, 1942, a veteran flying instructor and aircraft demonstrator named Betty Huyler Gillies, of Syosset, Long Island, received a telegram that read in part: AIR TRANSPORT COMMAND IS ESTABLISHING GROUP OF WOMEN PILOTS FOR DOMESTIC FERRYING STOP NECESSARY QUALIFICATIONS ARE COMMERCIAL LICENSE FIVE HUNDRED HOURS TWO HUNDRED HORSEPOWER RATING STOP ADVISE IF YOU ARE IMMEDIATELY AVAILABLE. The wire was signed by Henry H. "Hap" Arnold, commanding general of the United States Army Air Forces.

It was a message that many women fliers on both sides of the Atlantic would receive, in one form or another, during the course of World War II. The grim realities of global warfare led one country after another to reexamine the role of women in a time of national emergency. Women pilots in particular were valuable, for they already possessed at least the rudiments of a skill that was in desperately short supply. They represented, in the words of America's First Lady, Eleanor Roosevelt, "a weapon waiting to be used."

In England and America, the weapon was at first viewed with some alarm: "The menace," wrote C. G. Grey, the acerbic editor of Britain's highly regarded *The Aeroplane* magazine in 1939, "is the woman who thinks that she ought to be flying a high-speed bomber when she really has not the intelligence to scrub the floor of a hospital." In the United States, General Hap Arnold twice turned down proposals to use women for aircraft-ferrying duties, later explaining that he was not certain "whether a slip of a young girl could fight the controls of a B-17." In Canada, members of the Canadian Women's Auxiliary Air Force were denied flying responsibilities throughout the War and had to content themselves with replacing men in ground jobs instead.

Germany and Russia, by contrast, used their women pilots almost from the beginning. In Germany, women flew aircraft from the factories to fields near the battlefronts. Melitta Schiller and the renowned Hanna Reitsch even tested the most advanced Luftwaffe fighters.

German women remained civilians, but Russian women actually held military rank. Their military status made Russian women eligible for

A four-engined Stirling bomber dwarfs Joan Hughes, a ferry pilot in Britain's Air Transport Auxiliary, as she strides across an RAF airfield after delivering the winged behemoth from the factory.

combat, and they became the first women anywhere—apart from one Turkish female pilot in the 1930s *(page 137)*—to fly fighter and bomber missions. Of the other major combatant nations, neither Italy nor Japan used women fliers in its military program. The women of both countries—like Canadian and Australian women—were used for auxiliary services such as parachute rigging, aircraft assembly and clerical work. France did have a pool of highly skilled women pilots but was knocked out of the War before they could play any part.

The woman who opened the door to women pilots in Britain's war effort was a resolute and witty blonde named Pauline Gower, daughter of a distinguished member of Parliament, Sir Robert Gower. Pauline Gower had been co-owner of an air-taxi service during the early 1930s and was a commissioner of Britain's prewar Civil Air Guard, founded to train a reserve of pilots for emergency service. When she saw the Air Guard's wartime successor, the Air Transport Auxiliary (ATA), taking over ferrying responsibilities and rigorously excluding women, she began to use her influence to harry the Air Ministry.

The Ministry authorized her to form a women's section of the ATA, consisting of a mere eight recruits who were to ferry nothing larger than small training machines. The First Eight, as the press called them, were based at Hatfield in Hertfordshire, where they had to pay for their own lodgings, even though they were making 20 per cent less than male ferry pilots of the same grade. With at least 600 hours of flying experience apiece, they required no special training.

Throughout the wet, bitter winter of 1940, they ferried single-engined Tiger Moth trainers to distant fields in Scotland and the north of England. Their only protection from the weather as they droned northward was a small windshield across the open cockpit: "None of us will ever forget the pain of thawing out after such flights," Pauline Gower wrote later. Depending on weather conditions, the journey north might take several days to two weeks, she recalled—during which the pilot was forced to put in for rest and fuel at bleak RAF airfields that as yet had no facilities for women. Unable to sleep at the airfield, the women pilots usually hitched a ride to the nearest town to rent or beg a bed wherever they could find it.

To make matters worse, the ATA chief added, "we were, quite frankly, unpopular." To pilots of the RAF, it seemed that the ATA women were given far more press coverage than they deserved. Moreover, the male pilots were outspokenly dubious that the women were really competent. The only way to combat the "considerable prejudice" she encountered, Pauline Gower decided, was to compile a flight record that could not be criticized. "We didn't bend a blade of grass," she later noted proudly—pointing out that her recruits had no accidents at all during those first bleak months of 1940. (During the entire War, the accident rate of male and female ATA pilots was virtually the same.) The attitude of the women was succinctly expressed by an American pilot in the ATA, Mary Hooper: "It's two strikes out of three against women,

Turkey's tiny "Amazon of the Air"

Scarcely a decade after her countrywomen had been liberated from the harems of the ancient Ottoman Empire, a petite Turkish orphan named Sabiha Gökçen *(below)* became her nation's first woman flier, its first female Army pilot and probably the first woman anywhere to fly combat missions.

Influence helped. After her parents died in the upheavals that convulsed Turkey after World War I, Sabiha was adopted by Turkey's strong man, Kemal Ataturk, who ruled and modernized the country during the 1920s and 1930s. Finding his adopted daughter smitten by the rage to fly, Ataturk used his clout to have her admitted to a civilian flying school, then to the Turkish Army's air-cadet program.

The flier blazed to fame in 1937 when she joined a nine-plane force that helped to quell a revolt by Kurdish tribesmen. Called an Amazon of the Air by the press, she bombed and strafed the rebels for a month, emerging untouched although her slow plane was peppered by rifle fire. After this gust of glory she settled down to a prosaic job as a flight instructor. She continued to fly until 1966, logging 4,000 hours in the air.

Turkey's first woman pilot kisses the hand of the nation's leader, Kemal Ataturk, before joining her air-force unit in combat in 1937.

anyhow, so we are more conscientious and more determined not to have anybody say 'Just another woman pilot!' "

Led by Commander Pauline Gower (third from right), members of the Air Transport Auxiliary come home to the Hatfield air base on an ATA Anson taxiplane after a full day of ferrying aircraft to several delivery points about England.

For either men or women, ferrying airplanes in wartime Britain was not an easy task. Ferry pilots almost invariably had to follow a tortuous, zigzag route plotted to avoid barrage balloons, artillery ranges, RAF flying schools and coastal antiaircraft batteries. Denied, for security reasons, the use of radio, the fliers were forced to navigate by inadequate maps and sketchy knowledge of the countryside. Pilots had to cope with unexpected local navigational hazards in any way they could. One ATA woman coming in for a landing at an unfamiliar field in Wales slammed into a high-tension cable, cartwheeled her plane and miraculously emerged alive, though bleeding from facial cuts. "Shouldn't try a loop at 50 feet, Miss," said a flight engineer smugly. "It's too low for safety."

And ferry pilots ran the risk of an encounter with enemy aircraft. One

ATA pilots memorize landmarks of the English countryside before starting on their tortuous routes around barrage balloons, bad weather, artillery and RAF training fields. Wrote Commander Gower: "We have no such thing as a straight line."

of the most experienced ATA pilots, Margaret Fairweather, was just lifting off the runway one morning when she all but collided with a German plane coming in to bomb the field. She coolly wheeled out of machine-gun range and continued on her flight. As the ability of the ATA women to handle such emergencies became apparent, they were entrusted with bigger planes. Margaret Fairweather herself was soon flying the heaviest four-engined bombers. By 1943, women were ferrying virtually every kind of military plane in the British arsenal—120 different types of aircraft.

At the height of the War, more than 100 women were flying for the ATA. They included not only Englishwomen but Americans, Australians, Canadians, South Africans, New Zealanders, three Poles and a Chilean. The most famous recruit was long-distance flier Amy Johnson, who joined quietly and served unobtrusively.

Pilots sometimes were called on to ferry as many as four different types of aircraft in a single day. They began before dawn, checking the weather charts and barrage-balloon patterns at the airfield, studying the narrow air corridors that would take them shuttling between factories, maintenance units, training bases and squadrons. Then, after one last fast cup of tea, they were out to the airfield with assignment chits in hand and bulky parachutes slung over their shoulders.

After special instruction in different basic groups of aircraft, an ATA pilot was expected to be able to fly any plane assigned, and to fly it anywhere. But no pilot could be familiar with every type of plane in every class, and much of the time the fliers were forced to rely on a remarkable book called *Ferry Pilots' Notes*. In admirably brief space, the *Notes* gave the vital data about each aircraft, and pilots often would dip into it quickly while taxiing to the runway in an unfamiliar plane. Jackie Sorour—who at 20 was the youngest woman in the ATA—recalled long afterward the look of unalloyed terror on the face of an air marshal hitching a ride with her in a twin-engined Albemarle when she turned to him with a smile shortly after takeoff, explained that she had never flown the Albemarle before and began riffling through her *Pilots' Notes*. Fortunately, the flight passed smoothly and left the air marshal with new respect for the resourcefulness of ATA women.

When ATA pilots crashed—as inevitably, some did—it was usually because of the peculiar hazards of wartime flying, with fatigue playing an important role. All told, 15 women were killed flying for the ATA. Others barely escaped death. Diana Barnato, a veteran ATA pilot, was ferrying a high-speed Tempest fighter plane in 1944 when an explosion of undetermined cause blew the floor out from under her feet. Strapped securely in her seat but with her feet dangling over space, she brought the plane in to an emergency landing at the nearest field.

Even the most experienced pilots were at the mercy of the weather. Jackie Sorour recalled January 5, 1941, as a particularly bad day—cold, foggy, lashed with snow showers and rain. Flying a twin-engined Airspeed Oxford trainer, she took off from a South Wales airfield for Kidlington, near Oxford. Her route took her through mountainous country and then eastward along the Bristol Channel.

"I should have turned back, but valleys beckoned invitingly like tunnels," she later wrote. "I shot into one and peered ahead, but the other end was blocked with a wall of cloud." She pulled back on the stick, climbed steeply and at 4,000 feet burst through the cloud cover into blazing sunshine. She flew on over a blanket of clouds, not knowing if land or sea lay below. With her fuel supply fast dwindling, she had no choice but to go down: "The clouds embraced me like water around a stone," she wrote. "I descended slowly. Two thousand feet. Fifteen hundred. One thousand. Six hundred." Just as it seemed she must crash, the clouds broke and revealed the gray waters of the Bristol Channel and the dim line of the Somerset coast off to the east. Twenty minutes later, she was safe on the ground.

A far more famous ATA pilot was also flying to Kidlington that day. Amy Johnson had taken off from Blackpool on the Lancashire coast just before noon. Like Jackie Sorour, she was flying a twin-engined Oxford. The hour of her scheduled arrival passed and she became long overdue. At 3:30 in the afternoon, a parachute broke through the black cloud cover over the Thames estuary, which was not only off Amy Johnson's plotted course but roughly 100 miles beyond her destination. Shortly thereafter, a twin-engined Oxford hit the water.

The pilot plummeted into the storm-churned, icy water not far from the trawler H.M.S. *Haslemere,* which was on convoy duty. The *Haslemere* altered course and made for the bobbing flying helmet in the sea. Partway there, the trawler ran aground. Later there would be conflicting reports about what happened next. According to testimony of some of the *Haslemere's* crew, there were two figures struggling in the water that day—although the ATA adamantly insisted that Amy Johnson was flying alone. What is clear is that the downed flier called "Hurry, please hurry!" in a voice that seamen recognized as a woman's. They reached out to her as the seas carried her close to the *Haslemere,* but she disappeared under the trawler's stern and was not seen again. The second figure, if there was one, was also lost to the killing seas. Since no body was ever recovered and the crashed plane quickly sank, no positive identification of the drowned flier was possible. But two containers that floated free of the plane were identified as Amy Johnson's. To Jackie Sorour, looking back on that bleak January 5th, only one thing was clear: "Unfairly," she wrote, "all the luck in the sky was with me that day."

The loss of a flier of Amy Johnson's skill, experience and prominence was a particularly hard blow in the winter of 1941. The pilot shortage was becoming acute. When the United States made a commitment to supply Lockheed Hudson bombers to England as part of the Lend-Lease Act of March 1941, the British faced the problem of finding aviators capable of transatlantic flight.

Late that month, American aviator Jacqueline Cochran, whose spectacular record-breaking career had been temporarily interrupted by the War, fell into conversation with Hap Arnold at a Washington luncheon. The discussion turned to the problem of the Lockheed Hudson bombers, which were then being ferried by civilian pilots, since military pilots could not be spared for the job. Jacqueline Cochran saw an opportunity. She had been agitating without success for an organization of American women pilots who could fly in noncombat roles if America should go to war. Now she suggested that she might deliver a bomber to Britain and won General Arnold's support. The trip by such a famous flier would demonstrate for skeptics in the American military establishment that women fliers had a wartime role to play.

British ferrying officials were less than enthusiastic: They doubted the ability of a woman—even one as accomplished as Jackie Cochran—to fly a bomber across the Atlantic. She went over their heads to her old friend Lord Beaverbrook, British Minister of Supply. His influence got her invited to Montreal, the takeoff point for the Lockheed hops to England, for a flight test. During three days in Montreal she went through 60 takeoffs and landings and was cleared for transatlantic flight. The assignment had scarcely been announced when the male ferry pilots threatened to go on strike; to assign a woman to a man's job, they said, was to call their own competence in question. In the end, a compromise was worked out: Jackie Cochran flew the bomber across the Atlantic on June 17, 1941, but she had to relinquish the controls to her copilot, Captain Grafton Carlisle, on takeoff and landing.

It was a smooth flight, and after it Jackie did not stay long in England. She called on Pauline Gower, who briefed her on the work of the British women pilots. Then she returned to the United States to try to organize a similar group.

She knew that there was a large, untapped pool of possible recruits, including women on college campuses who had signed up for the Civilian Pilot Training Program—a project sponsored by the federal government and conducted by government-selected universities and flying schools. When she sent out letters to almost 3,000 women fliers, asking if they would be willing to enlist in a women's flying corps, the response

Three American pilots (left), new recruits in Britain's Air Transport Auxiliary, cross an airfield in 1942 with fellow ATA members. British Commander Pauline Gower is fourth from left and beside her is American aviator Jacqueline Cochran, who recruited her countrywomen for the unit.

was enthusiastic. A typical reply was that from Brooklyn College student Lydia Lindner: "I'll be available any time, anywhere."

But Jackie Cochran's plan to set up a women's ferrying division was turned down. In an emergency, General Arnold believed, America's women fliers could help most by serving as copilots on domestic airlines, thus freeing male pilots for combat. Since the United States was not yet facing such an emergency, he suggested in August 1941, perhaps Jackie Cochran should recruit a detachment of American women to serve in the British ATA.

She pounced on the idea. In order not to deplete the reserve of American women fliers in case they were needed later, she purposely kept the ATA detachment small, finally selecting 40 candidates from her file. One of them was 21-year-old Ann Wood of Brunswick, Maine, who later recalled, "I received the longest telegram I have ever seen—approximately two feet in length—from Jackie Cochran, detailing the qualifications needed to come to Britain." After an interview in New York, she went with other recruits to Montreal, where they were checked out by an ATA pilot. There the group was winnowed down to a final 25 who were sent on to England.

Nancy Love relaxes against the wing of a PT-19 trainer on the day she tested the first applicants to the Women's Auxiliary Ferrying Squadron, the aircraft-ferrying unit that she organized for the United States Army Air Forces in 1942.

Regardless of prior experience, all the women were required to work their way through the three-month ATA training program at Luton airfield, 30 miles north of London. Inevitably, there were complaints from the experienced American fliers that the training period was too long—although they stuck it out to the end. Jackie Cochran herself was criticized by some in the American contingent for accepting the honorary rank of flight captain although she necessarily spent more of her time administering than flying. Energetic and flamboyant, she was remembered long afterward for turning up at the flying field in a chauffeur-driven Rolls-Royce, dressed in her "second-best mink"—a display of riches that the British pilots resented at a time when the country was experiencing severe fuel and clothing rationing.

Nevertheless, even her critics acknowledged that she had recruited a first-class group of pilots. Most completed their 18-month contracts, and a few stayed for the duration of the War. Several were injured in the line of duty and one, Mary Nicholson, was killed when the propeller flew off the single-engined training plane she was ferrying.

Jackie Cochran returned home in the fall of 1942 confident that the time had come for the United States to mobilize its women fliers. When she arrived in New York and picked up a newspaper she was surprised to learn that the first steps in such a mobilization had already been taken. A Women's Auxiliary Ferrying Squadron, known as the WAFS, was being activated under the aegis of the Army Air Forces' Air Transport Command. The WAFS director was to be a woman named Nancy Harkness Love.

"Nancy Love was herself an exceedingly fine pilot," Jackie Cochran would acknowledge with considerable respect, "although I was reasonably sure that she did not go along with the idea of a large group of women pilots especially trained for various kinds of air work who would operate under military discipline."

She was right on both counts. Nancy Harkness Love, married to Major Robert Love, deputy chief of staff of the ATC's ferrying division, was a 27-year-old with 12 years of experience as a flier. Daughter of a Philadelphia physician, she had gone to school at Milton Academy in Massachusetts and spent her vacations building up flying hours. By the time she was 16, she had her private pilot's license. Two years later, when she was a Vassar student, she obtained her commercial and transport licenses. She and her husband owned an aviation company in Boston, for which she flew as a pilot. She also had considerable experience as a test pilot for the United States Bureau of Air Commerce. By the time she was appointed director of the WAFS, she had logged more than 1,200 hours of flying time.

Temperamentally, she was very different from the flamboyant Jackie Cochran. "It's stupid to call flying daredevilish," she once remarked. "I don't want to fly to the South Pole. I just want to do a job in the air. And I don't need to wear jodhpurs and fancy goggles to do it." Less aggressive than Jackie Cochran, she also had a more modest view of the wartime role women fliers should play. As early as May 1940 she had proposed recruiting a select group of highly qualified women pilots to supplement the all-male ferrying unit then being organized. While Jackie Cochran had been pushing for an entirely separate women's military program commanded not by men but by a woman—namely, herself—Nancy Love was interested only in integrating women into the Air Transport Command and leaving the direction to somebody else. Bored with administrative detail, she was happiest when she could get out of the office and into the cockpit of an airplane.

Nancy Love's proposal had been turned down in 1940, but it began to look more attractive after America got into the War at the end of 1941. By that time, women pilots all over the country were demanding a chance to serve. One letter that appeared in the *New York Herald Tribune* and attracted wide attention was from a young wife in Queens, New York. "Isn't there anything a girl of 23 years can do in the event our country goes to war, except to sit home and sew and become grey worrying?" she asked. "I learned to fly an airplane from a former World War ace, and I've forgotten how many parachute jumps I've made." She added: "If I were only a man there would be a place for me."

Nancy Love's WAFS would provide places for only a handful of America's thousands of eager women pilots. In the entire country, there were fewer than 100 women of the right age who could meet the program's rigid basic requirement of 500 hours in the air. In the fall of 1942, twenty-three of these women, averaging an impressive 1,100 hours of flying experience, were being checked out at the New Castle Army Air Base in Wilmington, Delaware.

Although they would be flying for the military, the women were hired as civilians. Originally, it was assumed that they would be given a 90-

Aviation history on a hangar wall

Aline Rhonie, a member both of America's Women's Auxiliary Ferrying Squadron and of Britain's Air Transport Auxiliary, was not only an experienced pilot but also an indefatigable artist who before the War had completed the largest painting on the subject of aviation ever seen in the United States.

Her gigantic mural—only a small portion of it is shown here—covered an entire 120-foot-long interior wall of a hangar at Long Island's old Roosevelt Field. Aiming to sum up flying's first two decades—"a thrilling era in aviation," as the artist said—she depicted 268 early planes as well as some 700 people who flew between 1908 and 1927, ranging from the Wright brothers to Charles Lindbergh. She included virtually all of America's pioneer women pilots.

She worked on the mural from 1935 to 1938, employing fresco techniques she learned after flying to Mexico City in 1934 to study with the famous mural painter Diego Rivera. In winter, she recalled, the hangar was so cold that "I had to paint wearing my fleece-lined flying suit from my open-cockpit days."

Before the hangar that contained the fresco was torn down in the 1960s to make way for a vast shopping center covering Roosevelt Field, an expert art restorer carefully peeled the mural off the wall, and it was put into storage.

Artist and pilot Aline Rhonie (right) leans on the scaffold she used while painting her 12-foot-high mural depicting aviation's early years. The segment below includes a smiling portrait of Harriet Quimby (front row, sixth from left) and next to her Matilde Moisant and Blanche Scott.

day trial period, just as male ferry pilots were, and then commissioned in the Army Air Forces and given flight pay. But Congress had made no provision for flight pay for women. Rather than delay the whole program, the Army Air Forces signed on the WAFS fliers as Civil Service employees, pending legislation giving them military status.

Jackie Cochran, meanwhile, was vigorously protesting the fact that precedence had been given to Nancy Love's program while her own proposals had been virtually ignored. She won her point: Washington decreed that there would be two programs. Nancy Love's elite WAFS would ferry planes for the Air Transport Command; at the same time, Jackie Cochran would organize a Women's Flying Training Detachment to polish novice pilots for eventual service in the WAFS.

The original WAFS roster included several distinguished aviators. Betty Gillies, the first woman to qualify, was a charter member of the Ninety-Nines who had worked for Grumman Aircraft and logged some 1,400 flying hours. Evelyn Sharp of Nebraska was a barnstormer who had flown almost 3,000 hours. Boston's Gertrude Meserve had taught hundreds of Harvard and MIT students to fly, while Teresa

A Women's Flying Training Detachment trainee solos in a PT-19, whose open cockpit at times proved hazardous. One woman fell out of her plane during a spin but was saved by her parachute.

James of Pittsburgh was an instructor accomplished at upside-down aerobatic maneuvers. Cornelia Fort of Nashville, Tennessee, had been a civilian aviation instructor at a field near Pearl Harbor when the Japanese attacked. In the air with a student on December 7, 1941, she had been obliged to seize the stick and veer sharply away when a plane bearing the Rising Sun insignia zoomed directly toward her with guns chattering.

The WAFS started flying Piper Cubs and PT-19s, but before long the women were ferrying anything the Army Air Forces wanted moved. By January 1943 they were uniformed in gray slacks and patch-pocket jackets and were split into four squadrons—one remaining at the New Castle base in Delaware and the others going to Dallas, Texas; Romulus, Michigan; and Long Beach, California. Jackie Cochran's Training Detachment, meantime, found a temporary home at Howard Hughes Field, the Houston municipal airport. A civilian aviation school on the airport grounds was selected to provide the training under Army supervision and discipline.

The first trainees to turn up were between 21 and 35 years old (the minimum age was later dropped to 18) and each had at least 200 hours of flying time. That requirement would be lowered in subsequent classes until finally it stood at only 35 hours. As word of the program got around, Jackie Cochran was deluged with more than 25,000 applications, some from other countries. A Canadian woman wrote: "Because we are fighting on the same side, I would like to ask you on bended knee if there will be any chance of accepting me. P.S. I am physically fit, have a college degree, can sing the *Star Spangled Banner,* have never been a Nazi spy and would gladly take out U.S. papers if only it were possible."

The early arrivals needed all the enthusiasm they could muster, for they were given little official encouragement. They had to report for duty at their own expense, were issued no uniforms and were expected to hunt up their own billets in private homes, boardinghouses, hotels and dismal auto courts. The only bright spot, recalled Marjorie Kumler, who was in the first class of 28, was that Jackie Cochran was on hand to greet them when they checked in for processing in November 1942 at Houston's Rice Hotel. The famous aviator, she remembered, was wearing a silver lapel pin shaped like a propeller with a large rosette diamond for a hub. While her recruits gathered around, Jackie Cochran made a brief speech of welcome. "If things don't run smoothly at first," she said, "just remember that you will have the honor and distinction of being the first women to be trained by the Army Air Forces. You are very badly needed."

Needed or not, they were skeptically received. "You may think that you are pretty hot pilots," the training center's commanding officer, Captain Paul C. Garrett, told them when they assembled before him. "I'd advise you to forget it. You're here to learn to fly the way the Army flies." For the recruits, that meant a day that began at dawn, when they were picked up by Army trucks and driven to Howard Hughes Field.

They divided their time between flight training and ground-school studies that included everything from meteorology and navigation to how to disassemble an engine. The trainees went aloft in a motley assortment of aging and poorly equipped civilian craft—"claptrap equipment," as Jackie Cochran called it. Having been issued no uniforms or flying suits, they turned up in blue jeans or cut-down men's trousers. In the late afternoon, they did an hour of calisthenics, then had their dinner and boarded the trucks again for the 11-mile drive into town. Nobody ever complained about the jolting ride, recalled Laurine Nielsen, who came from the Black Hills of South Dakota: "We were all so gung-ho; we sang every mile we rode, every day."

Howard Hughes Field was only a few miles away from Ellington Field, a vast base for training combat pilots. Inevitably, the practice area over Houston was jammed with high-speed planes maneuvering at various altitudes. For Margaret Kerr of Ada, Oklahoma, who had been flying for more than two years, some of the most frightening moments of her career came while practicing night landings at Houston. "We'd be stacked up at the four corners of the field," she recalled, "waiting for our turn to come into the traffic pattern. The only trouble was we often had some of the men getting tangled up in our traffic patterns, which is about as scary a thing as one likes to have on a night flight."

By May 1943, the Training Detachment had been shifted to Avenger Field at Sweetwater, Texas. Located some 40 miles west of Abilene in west Texas, Sweetwater was surrounded by scrub country, with little vegetation but mesquite trees and buffalo grass. Dust storms were frequent, and the summers brutal, with the temperature hovering around 100°. While the Houston operation had been makeshift, Sweetwater was all business and Army routine. The women lived in barracks, six to a room. Up at dawn to the sound of reveille, they marched to breakfast, to the flight line, to ground school and calisthenics. Their training was based on the regular Army course given to aviation cadets except for gunnery training and formation flying and certain fundamentals the women did not need because they already knew how to fly. When their 16-hour day ended, they went to bed to the sound of taps.

The military emphasis was due to the insistence of Jackie Cochran. Her recruits could not be formed into an efficient and reliable unit, she argued, unless they were subject to military discipline. There was plenty of that at Avenger Field, which was commanded by an Army major and administered by an Army staff—although the flight instructors were still civilians. The women trainees were expected to participate in parades and infantry drills, stand roll call and barracks inspections, and take the same oath of allegiance that was given to male cadets. They were not permitted to leave the base without a pass, and they received demerits for various infractions. Later, the women pilots would carry side arms and fly aircraft loaded with secret equipment restricted to all but military personnel.

It was an odd situation, since the women were in fact civilians. Al-

Sister Mary Aquinas (top), whom the press dubbed the Flying Nun for her interest in aviation, explains airplane engines to other nuns at Washington National Airport in 1943. The sisters then taught their high-school students about aviation in a Civil Aeronautics Authority program.

Enjoying a respite from their rigorous training at Avenger Field in Sweetwater, Texas, fledgling members of the Women Airforce Service Pilots (WASP) sunbathe between their barracks, only yards from their PT-19 trainers (background).

though they soon began to wear khaki pants, white shirts and khaki overseas caps to distinguish themselves, they did not have an official uniform of any kind until Jackie Cochran outfitted them with blue slacks, battle jackets and berets in the spring of 1944. But uniformed or not, they were expected to conform to military regulations as well as a rigid code of conduct that Jackie Cochran herself prescribed. "You had to get out of the airplane looking like a lady," recalled Joan Olmsted, who after the War became an Air Force intelligence officer. "Jacqueline Cochran had very rigid rules about how we should look and how we should conduct ourselves. But, God love her, it was good discipline."

Not all of the women felt the same way. "Some weren't all that thrilled with marching around in a Mickey Mouse army," Peggy Werber recalled. "A lot left because they didn't like it." In fact, there was nothing to prevent a woman from resigning at any time. As Dorothy Deane Ferguson pointed out, "We were volunteers: There was no one standing over us with a gun saying we had to stay." It is a measure of the women's determination that the rate of attrition of trainees in the women's program was about 35 per cent, virtually the same as the rate among male flying cadets—who did not have the option of resigning.

Jackie Cochran had been reasonably confident when she started the training program that women pilots would eventually be militarized and would be eligible for hospitalization, insurance, and the same veterans' and death benefits that other military personnel were entitled to. But Congress, with much other pressing business before it, failed to act until

Two amused WASP trainees model their baggy "zoot suits," the WASP nickname for the ill-fitting flying coveralls that the Army issued during the organization's early days. Later the shapelier official WASP uniform became available.

June 21, 1944. When a bill to militarize women pilots finally came to a vote on that date, it was defeated—largely because of a strong lobbying effort mounted by civilian flight instructors under contract to the Army. With victory in Europe in sight and the pilot shortage over, the Army Air Forces was closing down its special wartime pilot-training program—throwing some 10,000 highly paid civilian flight instructors out of jobs and making them eligible for the draft. These men felt they should be given jobs as ferrying pilots—and they feared that women pilots, if militarized, would automatically have priority for ferrying assignments. (More than three decades later, in 1977, Congress finally paid the nation's debt to the women fliers of World War II, retroactively militarizing them and making them eligible for veterans' benefits.)

Shortly after the move to Sweetwater, Jackie Cochran convinced the Army Air Forces that the training and operating branches of the women's pilot program—the WAFS ferry pilots and her own trainees—ought to be under one command, preferably her own. Accordingly, on August 5, 1943, the Sweetwater trainees and Nancy Love's WAFS were merged into one organization known as the Women Airforce Service Pilots, or WASP. Jackie Cochran became Director of Women Pilots, while Nancy Love became the WASP executive on the staff of the Air Transport Command's ferrying division.

By that time, many commanding officers preferred WASPs to male ferry pilots, because the women delivered their planes quicker. One possible reason, suggested General Arnold, was that the WASP "didn't carry an address book with her." A more probable reason was that women recruits simply threw themselves into their work with an enthusiastic dedication.

Fueled by their enthusiasm, the women pilots soon became involved in virtually every kind of flying except combat and overseas ferrying. But it still seemed, at times, that the more official accolades they received, the more they were resented. Betty Gillies, who was so tiny she had to have special wooden blocks fitted over the rudder pedals of many of the planes she flew, could understand how the male pilots must have felt when they saw a bantamweight woman climb down from a military plane. "The fighter pilots fresh out of school would let it be known how difficult the planes were to fly," she recalled. "They resented the girls flying those aircraft, and I can't blame them. I was kind of embarrassed about it."

Embarrassed or not, Betty Gillies flew every tough mission that came her way—as did the other women pilots. One of Jackie Cochran's first moves as chief of the WASP was to assign 50 of her pilots to tow targets for student antiaircraft gunners at Camp Davis, North Carolina. It was a job military pilots detested and avoided when they could. Standard procedure was for the pilot to cruise at about 10,000 feet over the Camp Davis dunes, towing a sleeve-shaped muslin target while the students fired guns as large as 90 millimeters at it. The tendency of novice gunners was to track too slowly and fall behind the target; urged by their officers to take a more generous lead, the gunners sometimes wound up shooting closer to the plane than the muslin sleeve, even though the tow rope was a minimum of 2,500 feet long. Every WASP who flew the hazardous miles knew the sensation of being bracketed by flak and fighting to keep the plane steady in a sky suddenly bumpy with explosions. Some returned to base with holes in the aircraft's tail section.

Neophyte gunners were not the only problem. The planes assigned to target work were old, war-weary and sometimes inclined to quit in mid-air. Fuel, in those days of severe rationing, was of such low octane that pilots were never confident their engines would start, or that they would keep running after takeoff. Everybody knew the fate of WASP Mabel Rawlinson, who was checking out an A-24 at Camp Davis one night when the plane inexplicably shuddered in the air, burst into flames and plummeted into a swamp near the field. Would-be rescuers, unable to approach the plane through the fire, could hear the pilot's screams until she burned to death.

The women's success in nonferrying assignments led the Army Air Forces command to take a far broader view of their capabilities. Henceforth, it was decided, they would be used for smoke laying during exercises, for engineering test flights, for simulated gas attacks and for night and day missions training radar and searchlight trackers. One assignment they particularly enjoyed was simulated low-level strafing while gun crews practiced tracking them. "Peeling off with the sun at our backs," recalled Winifred Wood of Miami, "we'd dive down on emplacements, trucks, chow lines, or anything visible. It was legalized buzzing and we loved it."

Nevertheless, ferrying remained the women pilots' chief function. One group began flying Bell Aviation's hot and tricky pursuit plane, the P-39, taking it from the factory at Niagara Falls to Great Falls, Montana. There the P-39s were picked up by male pilots who ferried them to Fairbanks, Alaska, to be turned over to the Russians as Lend-Lease aid. Other women pilots flew Republic's P-47 Thunderbolt from the factory at Farmingdale, Long Island, to Newark, New Jersey, for shipment to the European theater. Betty Gillies, for one, never got over her sense of what "a tremendous privilege" it was "to fly those beautiful aircraft." At the bomber-training school in Birmingham, Alabama, WASPs Dora Dougherty and Dorothea Johnson checked out in the B-29 Superfortress and found it "so beautifully balanced" they could hardly believe it. At the experimental flight center in Dayton, Ohio, a 21-year-old WASP named Ann Baumgartner streaked across Wright Field at 350 miles per hour in Bell Aircraft's top-secret, twin-turbine jet fighter, the YP-59A—America's first experimental jet.

At times a WASP might ferry 10 aircraft in a day. An unlucky pilot, on the other hand, might get involved in a string of unforeseen long-distance flights that would keep her on the move for several days or even weeks. Teresa James once took off from the Army Air Base at New

Castle, Delaware, with orders to pick up a P-47 at the Republic factory in Farmingdale and fly it to Republic's modification center at Evansville, Indiana. She expected to be gone a day; instead, she was gone for four hectic weeks, during which she crisscrossed the country as one plane after another was assigned to her at every successive base she touched.

Forbidden to fly after sundown, ferrying pilots had the nighttime responsibility of guarding their own planes at understaffed airfields. When flying the more advanced aircraft, they carried .45-caliber pistols to be used to protect the plane and its top-secret equipment if they were forced down under suspicious circumstances and feared the plane might fall into the hands of spies.

Women pilots compiled an enviable record. The WASPs delivered 12,650 planes of 77 different types. Fully 50 per cent of the ferrying of high-speed pursuit planes in the United States was done by WASPs. On these and other assignments, they flew a total of 60 million miles. Of the 1,830 women admitted to the WASP program, 1,074

A WASP flier tows a tattered muslin target that has been riddled by antiaircraft trainees practicing marksmanship from the ground. The sleeve was normally reeled out from the plane much farther than seen here, and no WASP was ever hit by flak, despite many close calls.

graduated and 38 lost their lives. None was more mourned than Evelyn Sharp, one of the most experienced of ferry pilots, who died when she was thrown through the canopy of her P-38 as she crash-landed near Harrisburg, Pennsylvania. Since she was a civilian and did not qualify for death benefits, her fellow pilots took up a collection to help pay for her funeral. Although they were officially not entitled to do so, the townspeople of her hometown of Ord, Nebraska, draped her coffin with an American flag.

On December 20, 1944, when women pilots were at the peak of their effectiveness, the WASP program was halted. Because combat losses had been far lower than anticipated, Army Air Forces pilots were now returning from overseas and taking over stateside flying duties formerly assigned to the WASPs. Although General Arnold was in favor of keeping women pilots active in some capacity, he could not justify doing so unless they were militarized. When a last effort to get a militarization bill through the Senate failed, he had no choice but to announce the end of the program. To most women pilots, the announcement was a jarring shock. Katherine Landry summed up the prevailing feeling in a succinct telegram to her family: CAN YOU USE A GOOD UPSTAIRS MAID WITH 800 FLYING HOURS?

The only place an American woman could still fly a military plane was in England. In the fall of 1944, the German armies were thrown back to the Rhine, and the British ATA began ferrying replacement aircraft for the RAF directly to the Continent. Women began ferrying planes across the Channel in January 1945. Among them were three Americans, Peggy Lennox, Nancy Miller and Ann Wood, who had come over with Jackie Cochran's original contingent and who continued flying with the ATA virtually until it closed down in November 1945.

On the war-wracked Continent, women would continue to fly until the last bitter battles were lost or won. Although there was no equivalent to the ATA or WASP in Germany, many individual women worked as test pilots or ferried warplanes to the various fronts. Vera von Bissing, Liesel Bach and Beate Uhse distinguished themselves flying Stukas, Focke-Wulfs, Messerschmitts and even a few Me 262 prototype jet fighters to the battle areas, particularly from the Atlantic to the Caspian Sea. Melitta Schiller received the Iron Cross and the diamond-studded Military Flight Badge for conducting a virtually unprecedented 1,500 test dives of German dive bombers. A graduate engineer, she analyzed the results of the tests herself.

The most famous woman pilot in wartime Germany was Hanna Reitsch, whose exploits for the military surpassed even her achievements as a glider and test pilot in the 1930s. For the Luftwaffe, she did virtually everything but fly combat missions. Aside from testing new fighters and bombers, she flew important Nazi leaders and military figures around the occupied territories and operational theaters, and she acted as liaison between advance bases and the rear echelon. She received the Iron Cross, Second Class, from Hitler's hand for testing a device that was designed to shear through the cables of the British barrage balloons over London. During one of the tests in Germany, a cable shaved off parts of her Dornier Do 17's propeller blades; she coolly feathered what remained of the propellers and brought the plane down to a safe landing.

Soon she was on an even more difficult mission—testing the German Me 163 experimental rocket plane. A viciously unstable machine, the Me 163 took off in a roar of flame and could soar to 30,000 feet in 90 seconds, eventually attaining a speed of nearly 600 miles per hour. Even sitting on the ground in "a hellish, flame-spewing din," she recalled, it was "all I could do to hold on as the machine rocked under a ceaseless succession of explosions." On her fifth flight, a special launching undercarriage failed to drop off as it was supposed to, and she was forced to come in for a crash landing with the plane bucking and rolling. Sitting stunned in the wrecked machine, she felt a stream of blood on her face and gingerly raised her hand to find that "at the place where my

Eight members of the Women Airforce Service Pilots undergo tests in an altitude chamber at Randolph Field, Texas, to measure their reactions to flying at 35,000 feet in an unpressurized cabin. After such tests the WASPs could be trained to ferry the high-flying B-17 bomber.

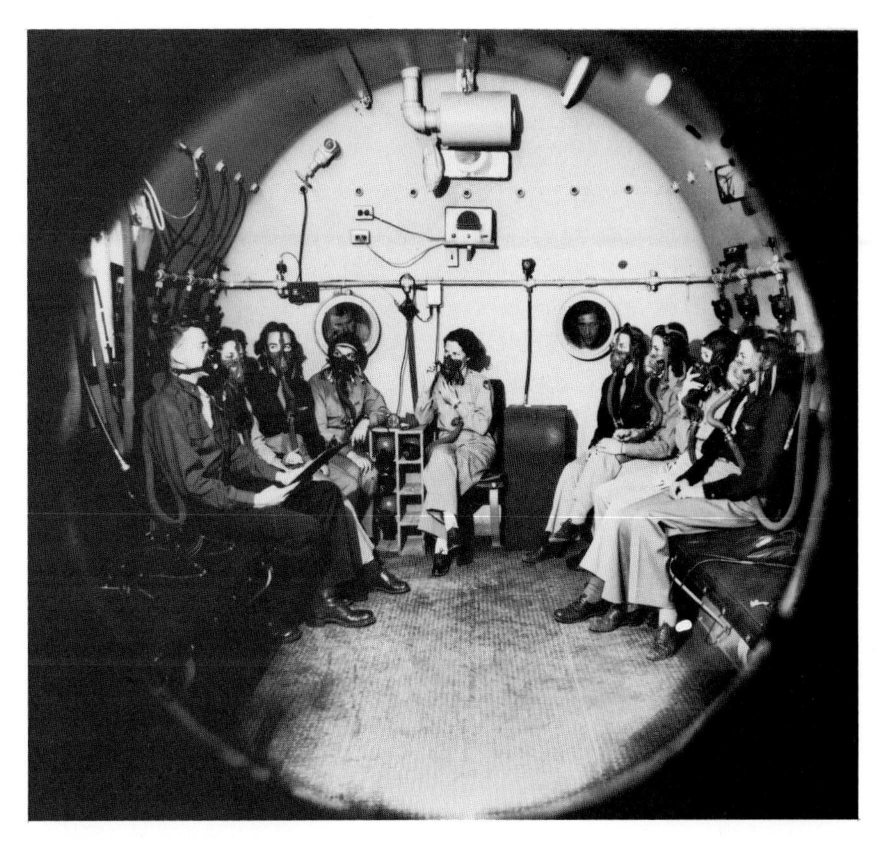

nose had been was now nothing but an open cleft." Always the professional, she reached for a pad and pencil and drew a sketch of the sequence of events leading to the crash. Then she blacked out.

She spent more than five months in the hospital recuperating from head injuries that her surgeons had thought would be fatal. Then, while still suffering from headaches and dizzy spells, she put herself through a strict regimen of tree and roof climbing to regain her sense of balance. Soon she was test-flying again, to the astonishment and concern of her doctors.

Worried about the progress of the War, she offered to form a squadron of women pilots "to fight for the Fatherland, on the same terms as the men of the Luftwaffe—without any privileges or restrictions." She was turned down but soon was active with plans to form a suicide squadron that was to strike at vital production centers in England and key warships of the Allied fleet. The aircraft to be used was the V 1 rocket, the famous buzz bomb that was then in the final stages of development. For the suicide project, the V 1 was provided with a seat for a pilot but with no landing gear, since this was to be a one-way flight. Hanna Reitsch successfully test-flew a prototype of the V 1 with special landing skids attached, and Hitler provisionally approved the project. It was abandoned when the Allies landed in Normandy in June of 1944.

Hanna Reitsch's most dangerous assignment was to ferry Luftwaffe General Robert Ritter von Greim to meet with Hitler in the bunker of the Berlin Reich Chancellery in the last days of the War. The Luftwaffe officers who briefed them before the flight clearly believed the mission was impossible: Berlin was completely surrounded by the Soviets, and for two days not a single German plane had been able to get into the city. Only one airport, Gatow, remained in German hands, and it was under continual artillery fire and expected to fall to the Soviets momentarily. Moreover, it was separated from the Reich Chancellery in the center of Berlin by 13 miles of Soviet-held territory.

Nevertheless, Hanna Reitsch and Greim decided to go ahead. The plan was to fly from Rechlin airport, 60 miles to the northwest of Berlin, in a Focke-Wulf Fw 190 single-seater fighter that had been modified to carry a passenger. A Luftwaffe sergeant with an excellent knowledge of Soviet antiaircraft defenses around Berlin would fly the plane to Gatow, and Hanna Reitsch would accompany Greim the rest of the way to the Chancellery. With the sergeant piloting the plane and Greim in the passenger seat, she had to be wedged feet-first into a space in the rear of the fuselage.

The virtually nonexistent Luftwaffe somehow mustered 30 or 40 fighters to escort the Focke-Wulf, and on April 26, 1945, the plane

Holding the Iron Cross, Second Class, she has just received, German heroine Hanna Reitsch is commended in 1941 by Adolf Hitler (right) and Luftwaffe chief Hermann Göring (left) for a dangerous flight test of a device for cutting the tethers of British barrage balloons.

managed to reach Gatow safely. For the flight to the Chancellery, Hanna Reitsch and Greim chose a Fieseler Storch. At the last moment, Greim decided to take the controls himself, since the woman aviator had no experience flying under fire. They got off the shell-pocked runway safely and flew through the outskirts of Berlin at treetop level to avoid Soviet fighters. Suddenly the fighters appeared and at the same time, recalled Hanna Reitsch, "from the ground, out of the shadows, from the treetops themselves, leapt the very fires of Hell." She looked down: "Below, Russian tanks and soldiers were swarming among the trees."

Greim was hit, and Hanna Reitsch seized the controls. Though the plane was riddled with armor-piercing shells, she somehow managed to land it near the Brandenburg Gate in Berlin's center. There the pair hitched a ride in a staff car and arrived at last at the Chancellery.

They stayed scarcely more than a day, during which Hanna Reitsch observed a Hitler who had almost completely lost touch with reality. He appointed General von Greim Supreme Commander of the Luftwaffe to replace Göring, who he felt had betrayed him. Like everybody else in the bunker, the new arrivals were given vials of poison to permit them "freedom of choice." Before they could use them, Hitler ordered Greim to leave and bring every available plane to the relief of Berlin. A low-wing Arado Ar 96 monoplane had managed to land near the Brandenburg Gate, and Hanna Reitsch and Greim were to fly it out of the besieged capital.

Shortly after midnight on April 28, SS troops accompanied them in an armored vehicle on an eerie ride through the ruined streets. While shells crashed nearby and Soviet searchlights probed the sky, Hanna Reitsch gunned the Arado down the smoky road and somehow got it into the air. Skimming the shattered rooftops, she headed out of the burning city. The searchlights followed her and so did the guns. "But miraculously," she later recalled, "not a single shot touched the plane." What finally saved her was a low cloud formation on the outskirts of the city. She plunged into it and emerged 12 miles later, free of both antiaircraft fire and Soviet fighter planes. They landed at Rechlin at three in the morning. Two days later they heard of Hitler's suicide.

Even in the last desperate days of the War Germany hesitated to recruit its women fliers for combat. In the Soviet Union, by contrast, women were flying combat missions almost from the beginning. Having suffered nearly catastrophic losses of pilots and planes during the German drive into the Ukraine in the summer of 1941, the Soviet high command called on one of the U.S.S.R.'s most experienced female aviators, Marina Raskova, to organize three regiments of women fliers. Most of the 200 recruits she selected were between 18 and 22 and came either from the civil air fleet or from the flying clubs that had sprung up all over the Soviet Union in the 1930s.

In October of 1941, Major Raskova began forming the women into the 586th Fighter Regiment, flying the Yakovlev YAK-1, and later the

YAK-3 and YAK-9 fighters; the 587th Bomber Regiment, flying Sukhoi/SU-2 short-range bombers; and the 588th Night Bomber Regiment, flying Polikarpov PO-2 biplanes. This last group later became famous under a new designation, the 46th Guards Night Bomber Regiment *(pages 160-171)*. In addition, there were women pilots who flew with largely male units. The most famous was Lilya Litvyak, who became a top woman ace by destroying 12 German planes and who fought at Stalingrad with the mostly male 73rd Fighter Regiment. Lieutenant Litvyak was shot down over Kharkov in 1943, at the age of 22. Another woman, Valentina Grizodubova, commanded the 300 male pilots and technicians of a long-range night bomber squadron.

Women's air-combat units served in the Ukraine, the northern Caucasus, the Taman Peninsula, the Crimea and at Stalingrad and eventually fought all the way to Berlin, bombing German railroad junctions, supply trains and ammunition dumps along the way. In two and a half years, the 586th Fighter Regiment fought in 125 separate air battles.

German fliers never got over their surprise at finding themselves in air battles with women. Luftwaffe Major D. B. Meyer recalled being attacked by a group of Soviet YAK-9s near Orel. The jettisoned canopy of his plane dropped into the propeller of a YAK, which immediately crashed. Meyer was astounded when he landed and approached the wreckage to find that the crumpled body in the cockpit was that of a woman, without rank insignia, identification or parachute.

On another occasion, in June of 1943, a group of Messerschmitts was flying air cover over German artillery emplacements at Krymskaya in the foothills of the Caucasus. Suddenly the pilots saw nine Soviet dive bombers approaching, flanked by a wing of fighters. As the Messerschmitts peeled off to attack, the Germans were amazed to hear female voices calling to one another on their radios. The Germans were drawn into a cross fire and in seconds lost four planes.

Examples of extreme courage among Soviet women pilots were almost the rule. In one celebrated case, two lone YAKs of the 586th Fighter Regiment intercepted 42 German bombers en route to strike the rail junction of Kastornoye in the central Russian uplands. Despite the odds, the two women pilots attacked immediately, shooting down four bombers before their own planes were disabled.

In 1943 Marina Raskova was killed in combat. To symbolize the debt it owed its women fliers, the Soviet government held the first state funeral since the beginning of the War, placing Major Raskova's ashes in the Kremlin wall with full military honors. Other combatant nations owed a similar debt to their women pilots. None but the Soviet women actually fought, but all of them demonstrated in countless ways that they deserved victory in the long battle for equality in the air. "It is on the record that women can fly as well as men," said General Arnold to the last graduating class of WASPs. He added that if there should be another national emergency, "we will not again look upon a women's flying organization as experimental."

Serenely self-confident, Valentina Grizodubova (center), who later commanded a Soviet long-range bomber squadron in World War II, sits among the students and staff of a flying school she directed in 1933. She went on to win a number of world records in aviation and became the head of a Soviet airline.

Glory and death for Russia's valiant women

In May 1942 a gruff Soviet division commander threw up his hands in dismay when he learned that a new regiment of combat pilots assigned to him consisted of 254 women, most of them less than 20 years old. Worse still, this 588th Night Bomber Regiment was to make night bombing raids on the stoutly defended enemy line north of the Crimea.

But as the 588th's aviators set to work pounding the enemy, the division commander's opinion of them improved remarkably. Flying through the Russian winter in the open cockpits of old canvas Polikarpov PO-2 biplanes that carried only four bombs and barely mustered 60 miles per hour, the women made 15 to 18 round trips a night through murderous German antiaircraft fire. Supported by highly skilled mechanics and bomb loaders, almost all of them women, they flew 24,000 sorties during the War and dropped 23,000 tons of bombs.

In the merciless air war over Russia, sex was no barrier to death, suffering or heroism. One night the Germans shot down four PO-2s within 15 minutes, killing eight women. Another night two women were shot down behind German lines but managed to walk back to their unit, dodging the enemy for three days. The women of the 588th once rescued a besieged Soviet garrison by airlifting food supplies through heavy German machine-gun fire for 40 days until their male comrades on the ground escaped.

In 1943 the 588th was awarded elite status with a new designation: the 46th Guards Regiment. By War's end the division commander had decorated every woman in the regiment, and 23 were honored with the coveted title Hero of the Soviet Union. Their unit was disbanded, and they returned to their factories and farms. "Even if we were to place at your feet all the flowers of the earth," eulogized an admiring group of male combat pilots, "they would not be a big enough tribute to your valor."

High-spirited women of the 46th Guards Regiment do a Russian folk dance between bombing missions in 1943.

Commander Yevdokia Bershanskaya briefs two women before they depart on a bombing raid; within weeks, both pilots were killed in combat.

Squadron leader Serafima Amosova, who was described by a comrade as "a superb flier, calculating and cool-headed," cleans her pistol after coming back from a bombing mission. She was promoted to second in command of the regiment for her leadership ability and prowess in the air.

The pilots push a truck out of the mud as the regiment moves to a new airfield within easier striking distance of the retreating Germans.

Fighting the bitter cold, mechanics (left) service a PO-2 hidden under snow-laden trees while a bomber pilot (right) waits to fly her mission.

Katya Ryabova (left), who flew 890 sorties, clasps comrade Nadya Popova as the two display their medals. In a single night they raided the Germans 18 times.

Radiant Katya Krasnokutskaya stands near her ambulance plane, a PO-2 with gondolas (right) fixed to the lower wings to carry the wounded.

Flying low to avoid detection, a formation of PO-2s skims across a Russian field on its way to bomb a German stronghold near the Crimea.

The regiment stands proudly at attention in 1943 as its division commander pins a medal, the Order of the Red Banner, on a navigator.

Acknowledgments

The index for this book was prepared by Gale Linck Partoyan. The editors also wish to thank Dora D. Strother and Fay Gillis Wells.
For their valuable help with the preparation of this volume, the editors wish to thank: **In Australia:** Canberra—Australian War Memorial; Miami—Mrs. Harry Bonney; Sydney—Margaret Kentley; Nancy Bird Walton. **In East Germany:** East Berlin—Hans Becker, Adn-Zentralbild. **In France:** Paris—Elisabeth Boselli; André Bénard, Odile Benoist, Elisabeth Bonhomme, Alain Degardin, Gilbert Deloizy, Georges Delaleau, Général Paul Dompnier, Deputy-Director, Yvan Kayser, Jean-Yves Lorant, Général Pierre Lissarague, Director, Stéphane Nicolaou, Pierre Wilefert, Curator, Musée de l'Air; Edmond Petit, Curator, Musée Air-France; Ste.-Adresse—Jacqueline Boucher. **In Great Britain:** Chipping Sodbury—E. J. Viles, M.B.E., ATA Association; Hendon—R. W. Mack, P. Merton, Alison Uppard, Royal Air Force Museum; Horne—Mrs. D. Barnato Walker; London—T. C. Charman, J. S. Lucas, J. O. Simmonds, M. J. Willis, Imperial War Museum; Alison King; Marjorie Willis, BBC Hulton Picture Library; Sywell—H. M. Newton, Northamptonshire Aero Club; Twyford—Lettice Curtis. **In Hong Kong:** Lawrence Chang; Linda Wu, Editor, *Echo Magazine.* **In Italy:** Florence—Professor Leonetto Tintori, Palazzo Pitti; Milan—Mario Grugnola; Rome—Countess Maria Fede Caproni, Museo Aeronautico Caproni di Taliedo. **In Japan:** Tokyo—The Japan Women's Association of Aeronautics. **In the Soviet Union:** Moscow—Photokhronika TASS. **In Spain:** Teneriffe—Jean Batten. **In Turkey:** Ankara—Sabiha Gökçen. **In the United States:** California—Academy of Motion Pictures Arts and Sciences; Elgen Long Enterprises, Betty Huyler Gillies; Lockheed Corporation; Eva McHenry; Betty E. Mitson; Bruce Reynolds, San Diego Aero-Space Museum; Ruth Rueckert; Yvonne Smith; Neta Snook Southern; Bobbi Trout; John Underwood; University of California, Los Angeles, Research Library; Washington, D.C.—Margaret Kerr Boylan; Jean Ross Howard; Jerry Kearns, Prints and Photographs Division, Library of Congress; Frank H. "Bud" Kelly; Philip Edwards, Von Hardesty, Mimi Scharf, Catherine Scott, Pete Suthard, National Air and Space Museum, Smithsonian Institution; Florida—Viola Gentry; Teresa James; Annette Gipson Way; Idaho—Gene Nora Jessen; Indiana—Keith Dowden, Assistant Director, Special Collections, Purdue University Libraries; Dorothy Niekamp; Kansas—Frank Pedroja, Beech Aircraft Corporation; Maryland—Patricia Thaden Frost; James L. Webb; Massachusetts—Frank Lavine, Director, Medford Public Library; Robin McElheny, The Schlesinger Library, Radcliffe College; Muriel Earhart Morrissey; Boston's Museum of Transportation; Bradford Washburn, Boston Science Museum; New York—Jeanna Alberga, Rockefeller Center Archives; Pamela Barker; Sue Bartczak; Margaret Werber Gilman; William K. Kaiser, Curator, Cradle of Aviation Museum; Sally Van Wagenen Keil; Leonard Donato, Allan Priaulx, Editor, King Features Syndicate; Elizabeth B. Mason, Associate Director, Columbia University Oral History Research Office, Butler Library; Cole Palen, Director, Old Rhinebeck Aerodrome; Doris Renniger, Manager, The Wings Club; Elinor Smith; Gordon Stone, Metropolitan Museum Costume Library; Ohio—Doris Scott, President, International Women's Air and Space Museum, Inc.; Virginia Thomas; Oklahoma—Loretta Jean Gragg, Executive Director, Lu Hollander, The 99s Inc., International Women Pilots Organization; Oregon—Laurine Nielsen; Pennsylvania—Henry A. Liese; Texas—G. Edward Rice, Curator, History of Aviation Collection, University of Texas at Dallas; Virginia—Dana Bell, USAF Still Photo Depository; Paul B. Bryce; Washington—Nancy Miller Livingston. **In West Germany:** Bochum—Dieter Knippschild, Ruhr-Universität, Sektion für Publizistik und Kommunikation; Bonn—Paul Metzger Stadarchiv und Wissenschaftliche Stadtbibliothek; Eschwege—Vera von Bissing; Frankfurt—Hans Deutsch, Deutscher Aero-Klub; Holzeim—Klara Schiller; Koblenz—Meinhard Nilges, Bundesarchiv Koblenz; Munich—Elly Beinhorn; Rudolf Heinrich, Deutsches Museum; Georg Pasewaldt; Wilhelm Sachsenberg; Angelika Eckert, Mutz Trense, Vereinigung Deutscher Pilotinnen; West Berlin—Heidi Klein, Roland Klemig, Bildarchiv Preussischer Kulturbesitz; Axel Schulz, Ullstein Bilderdienst.
Particularly useful sources of information and quotations were: *Amy Johnson* by Constance Babington-Smith, Collins, London, 1967; *Last Flight* by Amelia Earhart, Harcourt, Brace, 1937; *The Fun of It: Random Records of My Own Flying and of Women in Aviation* by Amelia Earhart, Brewer, Warren and Putnam, 1932; *Those Wonderful Women in Their Flying Machines; The Unknown Heroines of World War II* by Sally Van Wagenen Keil, Rawson, Wade Publishers, 1979.

Bibliography

Books

Babington-Smith, Constance, *Amy Johnson.* London: Collins, 1967.
Bacon, Gertrude, *Memories of Land and Sky.* London: Methuen & Co., 1928.
Batten, Jean:
My Life. London: George G. Harrap & Co., 1938.
Alone in the Sky. Shrewsbury: Airlife Publishing Company, 1979.
Beinhorn, Elly, *Flying Girl.* Henry Holt, no date.
Boase, Wendy, *The Sky's the Limit.* Macmillan, 1979.
Burke, John, *Winged Legend: The Story of Amelia Earhart.* G. P. Putnam's Sons, 1970.
Cochran, Jacqueline:
Final Report on Women Pilot Program. Army Air Forces, no date.
The Stars at Noon. Little, Brown, 1954.
Craven, Wesley Frank, and James Lea Cate, *The Army Air Forces in World War II.* University of Chicago Press, 1958.
Curtis, Lettice, *The Forgotten Pilots: A Story of the Air Transport Auxiliary 1939-45.* Henley-on-Thames: G. T. Foulis & Co., 1971.
Danishevsky, I., ed., *The Road of Battle and Glory.* Moscow: Foreign Languages Publishing House, no date.
Davis, Burke, *Amelia Earhart.* G. P. Putnam's Sons, 1972.
Dwiggins, Don:
Hollywood Pilot: The Biography of Paul Mantz. Doubleday, 1967.
They Flew the Bendix Race. Lippincott, 1965.
Earhart, Amelia:
The Fun of It. Brewer, Warren & Putnam, 1932.
Last Flight. Harcourt, Brace, 1937.
20 HRS. 40 MIN. G. P. Putnam's Sons, 1928.
Hamlen, Joseph R. *Flight Fever.* Doubleday, 1971.
Harris, Sherwood, *The First to Fly: Aviation's Pioneer Days.* Simon and Schuster, 1970.
Hatfield, D. D., *Los Angeles Aeronautics 1920-1929.* Northrop University Press, 1973.
Jablonski, Edward:
Atlantic Fever. Macmillan, 1972.
Ladybirds: Women in Aviation. Hawthorn Books, 1968.
Johnson, Amy, *Sky Roads of the World.* London: W. & R. Chambers, 1939.
Keil, Sally Van Wagenen, *Those Wonderful Women in Their Flying Machines.* Rawson, Wade Publishers, 1979.
King, Alison, *Golden Wings.* London: C. Arthur Pearson, 1956.
Lauwick, Hervé, *Heroines of the Sky.* London: Frederick Muller, 1960.
Markham, Beryl, *West with the Night.* Houghton Mifflin, 1942.
Moggridge, Jackie, *Woman Pilot.* London: Michael Joseph, 1957.
Montague, Richard, *Oceans, Poles and Airmen.* Random House, 1971.
Morrissey, Muriel Earhart, *Courage Is the Price.* McCormick-Armstrong Publishing Division, 1963.
Nichols, Ruth, *Wings for Life.* Lippincott, 1957.
Niekamp, Dorothy R., *Women and Flight, 1910-1978: An Annotated Bibliography.* Manuscript, The Amelia Earhart Research Scholar Program, 1980.
Oakes, Claudia M., *United States Women in Aviation through World War I.* Smithsonian Institution Press, 1978.
Planck, Charles E., *Women with Wings.* Harper & Brothers, 1942.
Putnam, George Palmer, *Soaring Wings: A Biography of Amelia Earhart.* Harcourt, Brace, 1939.
Reitsch, Hanna, *Flying is My Life.* G. P. Putnam's Sons, 1954.
Rolt, L. T. C., *The Aeronauts: A History of Ballooning 1783-1903.* Walker and Company, 1966.
Smith, Robert T., *Staggerwing!: Story of the Classic Beechcraft Biplane.* The Private Press of Robert Stephen Maney, 1967.
Southern, Neta Snook, *I Taught Amelia to Fly.* Vantage Press, 1974.
Strippel, Dick, *Amelia Earhart: The Myth and the Reality.* Exposition Press, 1972.
Strother, Dora Dougherty, *The W.A.S.P. Program: A Historical Synopsis.* Air Force Museum Research Division, 1971.

Thaden, Louise McPhetridge, *High, Wide and Frightened.* Stackpole Sons, 1938.
Trout, Bobbi, *Just Plane Crazy.* Manuscript, no date.
Underwood, John W., *The Stinsons.* Heritage Press, 1969.
Wood, Winifred, *We Were Wasps.* Glade House, 1945.

Periodicals
"The All-Women's Meeting." *The Aeroplane,* September 23, 1931.
"An American Girl the First to Fly the English Channel." *American-Examiner,* 1912.
Beinhorn, Elly, "Marga von Etzdorf-allein nach Nordafrika und weiter." *Flieger-Kalender,* 1980.
Blanch, Lesley, "Pilots in Nailpolish." *Pegasus,* October 1943.
Blizstein, Madelin, "How Women Flyers Fight Russia's Air War." *Aviation,* July 1944.
Burtnett, Gerald B., "America's First Flying Sportswoman." *The Sportsman Pilot,* June 1931.
Clinton, Audrey, "Artist Hopes to Save Aviation Mural." *Newsday,* August 29, 1960.
Delear, Frank J., "What Killed Harriet Quimby?" *Yankee,* September 1979.
Earhart, Amelia, "My Flight From Hawaii." *National Geographic,* May 1935.
"The England-Australia Record." *The Aeroplane,* October 14, 1936.
Flight: September 25, 1931; October 22, 1936; October 28, 1937.
Flying: December 1916, January 1917, September 1942, December 1942, August 1943, January 1944, December 1944, March 1957.
Fort, Cornelia, "At the Twilight's Last Gleaming." *Woman's Home Companion,* July 1943.
Grahame-White, Claude, and Rudolph Hensingmuller, "Will Woman Drive Man Out of the Sky?" *American-Examiner,* 1911.
Johnson, Ann R., "The WASP of World War II." *Aerospace Historian,* Summer/Fall 1970.
Journal of American Aviation Historical Society, Winter 1964, Winter 1969, Fall 1979.
King, Anita, "Brave Bessie: First Black Pilot," *Essence,* May, June 1976.
Knight, Charlotte, "Our Women Pilots." *Air Force,* September 1943.
Kumler, Marjorie, "They've Done it Again!" *Ladies Home Journal,* March 1944.
Leslie's Weekly: May 25, 1911; August 17, 1911; August 24, 1911; May 16, 1912; June 6, 1912; July 18, 1912.
Marvingt, Marie, "Comment J'ai Concu le Premier Avion Sanitaire?" *Revue de la Société Scientifique et Historique de Documentation Aérienne,* January-February 1940.
"Miss Jean Batten's Great Achievement." *Speed,* November 1936.
Mitroshenkov, V., "They Were First." *Soviet Military Review,* March 1969.
Newcomb, Harold, "Cochran's Convent." *Airman,* May 1977.
The New York Times: November 28, 1909; June 1, 1930; September 5, 1930; June 21, 1932; January 15, 1933.
Quimby, Harriet:
"American Bird Women." *Good Housekeeping Magazine,* September 1912.
"How I Made My First Big Flight Abroad." *Fly Magazine,* June 1912.
"Ruth Law and Her Remarkable Flight from Chicago to New York." *Scientific American,* December 2, 1916.
"San Antonio Aviatrix Captures Japan; Katherine Stinson Thrills Nipponese." *San Antonio Express,* January 28, 1917.
Smith, Elinor, "This Business of Flying." *Liberty,* August 9, 1930.
Smith, Helena Huntington, "Profiles: New Woman, Elinor Smith." *New Yorker,* May 10, 1930.
Stinson, Marjorie, "Wings for War Birds." *Liberty,* December 28, 1929.
The Sywell Windstocking, October 1931.
Talbot, John F., "The Career of Matilde Moisant." *Popular Aviation,* February 1929.
Thaden, Louise, "The National Women's Air Derby." *Aviation Quarterly,* 1974.
Time, August 24, 1936; September 14, 1936.
Weishimer, Helen, "The Women Who Mark the Air Lanes." *EveryWeek Magazine,* July 17-18, 1935.
Zabavskaya, L., "Women Fighter Pilots." *Soviet Military Review,* March 1977.

Picture credits

The sources for the illustrations in this book are listed below. Credits from left to right are separated by semicolons, from top to bottom by dashes. Endpaper (and cover detail, regular edition): Painting by Paul Lengellé. 6, 7: Brown Brothers. 8: Smithsonian Institution Photo No. A 43515. 10: Henry Beville, courtesy National Air and Space Museum, Smithsonian Institution. 11: Culver Pictures. 12: Library of Congress. 14: Photo Bibliothèque Nationale, Paris. 16, 17: Musée de l'Air, Paris. 19: Museo Aeronautico Caproni di Taliedo, Rome—Suddeutscher Verlag, Bilderdienst, Munich. 20: Popperfoto, London—Carruthers Aviation Collection, Claremont Colleges; Fotokhronika TASS, Moscow. 21: Courtesy Bowen Bryant, Sydney; from Hsin Yi-Yun's Album, Taiwan—Yae Nozoki, Tokyo. 24: Elisabeth Hiatt Gregory Collection, UCLA. 25: Smithsonian Institution Photo No. 80-2022. 26, 27: Ullstein Bilderdienst, Berlin (West). 29: The Bettmann Archive. 30: Dmitri Kessel, courtesy Musée de l'Air, Paris. 32: Henry Beville, courtesy National Air and Space Museum, Smithsonian Institution. 34, 35: John W. Underwood Collection. 36: Library of Congress—John W. Underwood Collection. 37: Culver Pictures. 38, 39: John W. Underwood Collection; Culver Pictures. 40, 41: John W. Underwood Collection. 42: A. Y. Owen, courtesy The 99s Inc., International Women Pilots Organization, Oklahoma City. 45: Keystone Press, New York. 46: By permission of the British Library, Newspaper Division. 47: American Heritage Center, University of Wyoming. 49: Courtesy Purdue University Libraries—Schlesinger Library, Radcliffe College. 50, 51: Schlesinger Library, Radcliffe College, except second from left, bottom, Smith College Archives. 53: Central Press Photos, Ltd., London—The Bettmann Archive; Henry Groskinsky, courtesy Purdue University Libraries. 54: The Bettmann Archive. 56, 57: Courtesy Bill and Pat Thaden. 58: A. Y. Owen, courtesy The 99s Inc., International Women Pilots Organization, Oklahoma City. 60: Library of Congress. 61: Courtesy R. J. Reynolds Tobacco Company—Henry Beville, courtesy National Air and Space Museum, Smithsonian Institution. 62: A. Y. Owen, courtesy The 99s Inc., International Women Pilots Organization, Oklahoma City. 63: Wide World. 66, 67: Schlesinger Library, Radcliffe College. 68, 69: John W. Underwood Collection. 70, 71: John McDermott, courtesy Bobbi Trout (2); The Bettmann Archive (2). 72, 73: John McDermott, courtesy Bobbi Trout; courtesy Harcourt Brace Jovanovich, Inc. 74: Aldo Durrazzi, courtesy Aeronautico Caproni di Taliedo, Rome. 76: Painting by Cowan Dobson, courtesy Naval and Military Club, London. 78, 79: Derek Bayes, courtesy The Society of British Aviation Artists. 82: BBC Hulton Picture Library, London. 83: Freelance Photographers Guild. 86-91: Royal Air Force Museum, Hendon. 92, 93: John W. Underwood Collection. 95: Suddeutscher Verlag, Bilderdienst, Munich. 96: Ullstein Bilderdienst, Berlin (West). 97: Suddeutscher Verlag, Bilderdienst, Munich. 99: ADN-Zentralbild, Berlin (DDR). 100, 101: Ullstein Bilderdienst, Berlin (West). 102, 103: Courtesy Northamptonshire Aero Club; Central Press Photos Ltd., London (2). 104, 105: Central Press Photos Ltd., London. 106, 107: Courtesy Northamptonshire Aero Club. 108, 109: Clyde H. Sunderland, courtesy Purdue University Libraries. 110: Department of Transportation, Federal Aviation Administration. 113: A. Y. Owen, courtesy The 99s Inc., International Women Pilots Organization,Oklahoma City, and *The Pittsburgh Post Gazette.* 115: Courtesy National Portrait Gallery, Smithsonian Institution. 116, 117: Mark Sexton, courtesy Medford, Massachusetts, Public Library; A. Y. Owen, courtesy The 99s Inc., International Women Pilots Organization, Oklahoma City (2)—Henry Groskinsky, courtesy Purdue University Libraries (4). 120, 121: Drawing by John Amendola; courtesy Bill and Pat Thaden. 122, 123: Wide World. 124: Map by Walter Roberts. 126: Henry Beville, courtesy National Air and Space Museum, Smithsonian Institution—courtesy Purdue University Libraries. 127: Courtesy Purdue University Libraries. 129: Henry Beville, courtesy National Air and Space Museum, Smithsonian Institution. 130, 131: Drawing by John Batchelor. 132: Bradd Collection. 134, 135: Fox Photos Ltd., London. 137: Publifoto Notizie, Milan. 138: Fox Photos Ltd., London. 139: Courtesy Alison King, London. 142: Keystone Press, London. 143: Henry Beville, courtesy National Air and Space Museum, Smithsonian Institution. 144, 145: Smithsonian Institution Photo No. A-42046A—Cradle of Aviation Museum, Garden City, N.Y. 146: Peter Stackpole for *Life.* 148: National Archives Photo No. 208-PV-6a-1. 150, 151: U.S. Air Force Photos. 153: Courtesy Dora D. Strother. 155: U.S. Air Force Photo. 156: Bildarchiv Preussischer Kulturbesitz, Berlin (West). 158-171: Yevgeni Khaldei, Moscow.

Index

Numerals in italics indicate an illustration of the subject mentioned.